ORDER NUMBER EA-FAA-T-8080-11AX

AVIATION MECHANIC
POWERPLANT
QUESTION BOOK

1986

International Standard Book Number 0-89100-284-7
For sale by: International Aviation Publishers, Inc.
P.O. Box 36   1000 College View Drive
Riverton, Wyoming 82501-0036
Tel: 1 (800) 443-9250
(307) 856-1582

International Aviation Publishers, Inc.
1000 College View Drive, Riverton, Wyoming 82501-0036

CONTENTS

PREFACE

This question book has been developed by the FAA (Federal Aviation Administration) for use by applicants who are preparing for the Aviation Mechanic Powerplant Written Test and also for use by FAA Testing Centers and FAA Designated Written Test Examiners in administering the test. It is issued as FAA–T–8080–11A, Aviation Mechanic Powerplant Question Book, and is available to the public from:

> Superintendent of Documents
> U.S. Government Printing Office
> Washington, D.C.   20402

or from U.S. Government Printing Office Bookstores located in major cities throughout the U.S.

The questions included in this publication are predicated on regulations, principles, and practices that were valid at the time of publication. The question selection sheets prepared for use with this question book are security items and are revised at frequent intervals.

The Federal Aviation Administration does not supply the correct answers to questions included in this book. Students should determine the answers by research and study, by working with instructors, or by attending ground schools. The Federal Aviation Administration is NOT responsible for either the content of commercial reprints of this book or the accuracy of the answers they may list.

Comments regarding this publication should be directed to:

> U.S. Department of Transportation
> Federal Aviation Administration
> Aviation Standards National Field Office
> Examinations Standards Branch
> Airworthiness Section, AVN–133
> P.O. Box 25082
> Oklahoma City, Oklahoma   73125

REFERENCE GUIDE

ITP GENERAL . . . . AVIATION TECHNICIAN INTEGRATED TRAINING PROGRAM,
AVIATION MAINTENANCE PUBLISHERS

ITP AIRFRAME. . . . . AVIATION TECHNICIAN INTEGRATED TRAINING PROGRAM,
AVIATION MAINTENANCE PUBLISHERS

ITP POWERPLANT . . . . AVIATION TECHNICIAN INTEGRATED TRAINING PROGRAM,
AVIATION MAINTENANCE PUBLISHERS

AIRCRAFT ELECTRICITY AND ELECTRONICS, THIRD EDITION, McGRAW HILL,

AIRCRAFT MAINTENANCE AND REPAIR, FOURTH EDITION, McGRAW HILL

AIRCRAFT POWERPLANTS, FOURTH EDITION, McGRAW HILL

POWERPLANTS FOR AEROSPACE VEHICLES, THIRD EDITION, McGRAW HILL

AIRCRAFT GAS TURBINE ENGINE TECHNOLOGY, McGRAW HILL

AIRCRAFT PROPULSION POWERPLANTS, EDUCATIONAL PUBLISHERS, INC.

THE AVIATION MECHANIC'S MANUAL, McGRAW HILL

AC39-7A . . . . AIRWORTHINESS DIRECTIVES FOR GENERAL AVIATION AIRCRAFT

AC43-4 . . . . FAA ADVISORY CIRCULAR, CORROSION CONTROL

AC43-9A . . . . MAINTENANCE RECORDS: GENERAL AVIATION AIRCRAFT

AC43.13-1A. . . . . ACCEPTABLE METHODS, TECHNIQUES, AND PRACTICES,
AIRCRAFT INSPECTION AND REPAIR

AC43.13-2A. . . . . AIRCRAFT ALTERATIONS

AC65-9A . . . . AIRFRAME AND POWERPLANT MECHANICS GENERAL
HANDBOOK, U.S. DEPT. OF TRANSPORTATION

AC65-12 . . . . AIRFRAME AND POWERPLANT MECHANICS POWERPLANT
HANDBOOK, U.S. DEPT. OF TRANSPORTATION

AC65-12A . . . . AIRFRAME AND POWERPLANT MECHANICS POWERPLANT
HANDBOOK, U.S. DEPT. OF TRANSPORTATION

AC65-15A . . . . AIRFRAME AND POWERPLANT MECHANICS AIRFRAME
HANDBOOK, U.S. DEPT. OF TRANSPORTATION

AC65-19B . . . . INSPECTION AUTHORIZATION STUDY GUIDE

AC91-44A . . . . OPERATIONAL AND MAINTENANCE PRACTICES
FOR EMERGENCY LOCATOR TRANSMITTERS AND RECEIVERS

EA-FAR-1I . . . . FEDERAL AVIATION REGULATIONS, HANDBOOK FOR AVIATION
MECHANICS, AVIATION MAINTENANCE PUBLISHERS

EA-AAC-1 . . . . AIRCRAFT AIRCONDITIONING (VAPOR CYCLE),
AVIATION MAINTENANCE PUBLISHERS

EA-AC 61-13B . . . . BASIC HELICOPTER HANDBOOK, U.S. DEPT. OF TRANSPORTATION

EA-AGV. . . . . AIRCRAFT GOVERNORS,
AVIATION MAINTENANCE PUBLISHERS

EA-AH-1 . . . . AIRCRAFT HYDRAULIC SYSTEMS,
AVIATION MAINTENANCE PUBLISHERS

EA-AOS-1 . . . . AIRCRAFT OXYGEN SYSTEMS,
AVIATION MAINTENANCE PUBLISHERS

EA-APC. . . . . AIRCRAFT PROPELLERS AND CONTROLS,
AVIATION MAINTENANCE PUBLISHERS

EA-FMS-1 . . . . AIRCRAFT FUEL METERING SYSTEMS,
AVIATION MAINTENANCE PUBLISHERS

EA-TEP . . . . AIRCRAFT GAS TURBINE POWERPLANTS,
AVIATION MAINTENANCE PUBLISHERS

AIRCRAFT TECHNICAL DICTIONARY, SECOND EDITION

MAXIMUM TIME ALLOWED FOR TEST: 4 HOURS

Materials to be used with this question book when used for airman testing:

1. Airman Written Test Application which includes the answer sheet.
2. Question Selection Sheet which identifies the questions to be answered.
3. Plastic overlay sheet which can be placed over electrical drawings, graphs, and charts for making pencil marks or for tracing schematics for analysis purposes. If you are not provided the overlay, request one from the test monitor.

GENERAL INSTRUCTIONS

1. Read the instructions on page 1 of the Airman Written Test Application and complete the form on page 4.

2. The question numbers in this question book are numbered consecutively beginning with No. 4001. Refer to the question selection sheet to determine which questions to answer.

3. For each item on the answer sheet, find the appropriate question in the book.

4. Mark your answer in the space provided for that item on the answer sheet.

5. Remember:

    Read each question carefully and avoid hasty assumptions. Do not answer until you understand the question. Do not spend too much time on any one question. Answer all of the questions that you readily know and then reconsider those you find difficult.

If a regulation is changed after this question book is printed, you will receive credit until the affected questions are revised.

THE MINIMUM PASSING GRADE IS 70.

WARNING

§ 65.18 Written tests: cheating or other un-authorized conduct.

(a) Except as authorized by the Administrator, no person may—

(1) Copy, or intentionally remove, a written test under this Part;

(2) Give to another, or receive from another, any part or copy of that test;

(3) Give help on that test to, or receive help on that test from, any person during the period that test is being given;

(4) Take any part of that test in behalf of another person;

(5) Use any material or aid during the period that test is being given; or

(6) Intentionally cause, assist, or participate in any act prohibited by this paragraph.

(b) No person who commits an act prohibited by paragraph (a) of this section is eligible for any airman or ground instructor certificate or rating under this chapter for a period of one year after the date of that act. In addition, the commission of that act is a basis for suspending or revoking any airman or ground instructor certificate or rating held by that person.

AVIATION MECHANIC POWERPLANT

INTRODUCTION

The requirements for a mechanic certificate and ratings, and the privileges, limitations, and general operating rules for certificated mechanics are prescribed in Federal Aviation Regulations Part 65, Certification: Airmen Other Than Flight Crewmembers. Any person who applies and meets the requirements is entitled to a mechanic certificate.

At FAA Testing Centers, or at an FAA Designated Written Test Examiner's facility, the applicant is issued a "clean copy" of this question book, an appropriate 100–item question selection sheet which indicates the specific questions to be answered, and AC Form 8080–3, Airman Written Test Application, which contains the answer sheet. The question book contains all the supplementary material required to answer the questions. Supplementary material, such as an illustration, will normally be found within one page of the question with which it is associated. Where this is not practicable, page reference numbers will be given.

THE WRITTEN TEST

Questions and Scoring

The test questions are of the multiple–choice type. Answers to questions listed on the question selection sheet should be marked on the answer sheet of AC Form 8080–3, Airman Written Test Application. Directions should be read carefully before beginning the test. Incomplete or erroneous personal information entered on this form delays the scoring process.

The answer sheet is sent to the Mike Monroney Aeronautical Center in Oklahoma City where it is scored by a computer. Shortly thereafter AC Form 8080–2, Airman Written Test Report, is sent to the applicant listing the score and test questions missed.

The applicant must present this report for an oral and practical test, or for retesting in the event of written test failure.

Taking the Test

The test may be taken at FAA Testing Centers, FAA Written Test Examiners' facilities, or other designated places. After completing the test, the applicant must surrender the issued question book, question selection sheet, answer sheet, and any papers used for computations or notations to the monitor before leaving the test room.

When taking the test, the applicant should keep the following points in mind:

1. Answer each question in accordance with the latest regulations and procedures.
2. Read each question carefully before looking at the possible answers. You should clearly understand the problem before attempting to solve it.
3. After formulating an answer, determine which of the alternatives most nearly corresponds with that answer. The answer chosen should completely resolve the problem.
4. From the answers given, it may appear that there is more than one possible answer; however, there is only one answer that is correct and complete. The other answers are either incomplete or are derived from popular misconceptions.
5. If a certain question is difficult for you, it is best to proceed to other questions. After the less difficult questions have been answered, return to those which gave you difficulty. Be sure to indicate on the question selection sheet the questions to which you wish to return.
6. When solving a computer problem, select the answer nearest your solution. The problem has been checked with various types of computers; therefore, if you have solved it correctly, your answer will be closer to the correct answer than to any of the other choices.
7. To aid in scoring, enter personal data in the appropriate spaces on the test answer sheet in a complete and legible manner. Enter the test number printed on the question selection sheet.

Retesting—FAR Section 65.19

Applicants who receive a failing grade may apply for retesting by presenting their AC Form 8080–2, Airman Written Test Report—

(1) after 30 days from the date the applicant failed the test; or
(2) the applicant may apply for retesting before the 30 days have expired upon presenting a signed statement from an airman holding the certificate and rating sought by the applicant, certifying that the airman has given the applicant additional instruction in each of the subjects failed and that the airman considers the applicant ready for retesting.

NOTE: Blank spaces may appear on several pages of this publication. This permits referenced figures and tables to appear as close as possible to their related questions.

AVIATION MECHANIC POWERPLANT TEST—SECTION 1

4001. Which of the following statements is true regarding bearings used in high–powered reciprocating aircraft engines?

1—The outer race of a single–row, self–aligning ball bearing will always have a radius equal to the radius of the balls.
2—There is less rolling friction when ball bearings are used than when roller bearings are employed.
3—Crankshaft bearings are generally of the ball type due to their ability to withstand extreme loads without overheating.
4—Crankshaft bearings are generally of the ball type due to their ability to withstand loads. However, some manufacturers object to their use because this type bearing requires a positive high–pressure oil supply.

4002. Which of the following propeller reduction gear ratios will cause the highest propeller RPM? (Assume the same engine RPM in each case.)

1—16:7.
2—16:9.
3—20:9.
4—3:2.

4003. Which of the following indications would be the least likely to be caused by failed or failing engine bearings?

1—Excessive oil consumption.
2—High oil temperatures.
3—Low oil temperatures.
4—Low oil pressure.

4004. What is the principle advantage of using propeller reduction gears?

1—To enable the propeller RPM to be increased without an accompanying increase in engine RPM.
2—The diameter and blade area of the propeller can be increased.
3—To enable the engine RPM to be increased with an accompanying increase in power and allow the propeller to remain at a lower, more efficient RPM.
4—To enable the engine RPM to be increased with an accompanying increase in propeller RPM.

4005. Which of the following will decrease volumetric efficiency of a reciprocating engine?

1—High fuel octane rating.
2—Short intake pipes of large diameter.
3—Low carburetor air temperature.
4—High cylinder head temperature.

4006. Which of the following is a characteristic of a thrust bearing used in most radial engines?

1—Tapered roller.
2—Double–row ball.
3—Double–row straight roller.
4—Deep–groove ball.

4007. Which of the following bearings is least likely to be a roller or ball bearing?

1—Rocker arm bearing (overhead valve engine).
2—Master rod bearing (radial engine).
3—Crankshaft main bearing (radial engine).
4—Generator armature bearing.

4008. The horsepower developed in the cylinders of a reciprocating engine is known as the

1—shaft horsepower.
2—indicated horsepower.
3—brake horsepower.
4—thrust horsepower.

4009. A nine–cylinder engine with a bore of 5.5 inches and a stroke of 6 inches will have a total piston displacement of

1—740 cubic inches.
2—1,425 cubic inches.
3—23,758 cubic inches.
4—1,283 cubic inches.

4010. The five events of a four–stroke cycle engine in the order of their occurrence are

1—intake, ignition, compression, power, exhaust.
2—intake, power, compression, ignition, exhaust.
3—intake, compression, ignition, power, exhaust.
4—intake, ignition, power, compression, exhaust.

4011. If fuel/air ratio is proper and ignition timing is correct, the combustion process should

1—be completed 20° to 30° before top center at the end of the compression stroke.
2—be completed when the exhaust valve opens at the end of the power stroke.
3—continue until the end of the exhaust stroke.
4—be completed just after top center at the beginning of the power stroke.

4012. The clearance between the rocker arm and the valve tip affects how many of the following?

(1) Point at which valve opens.
(2) Height of valve opening.
(3) Duration of valve opening.

1—One.
2—Two.
3—Three.
4—None.

4013. Which statement is correct regarding engine crankshafts?

1—Counterweights reduce torsional vibrations.
2—Counterweights provide static balance.
3—A six–throw crankshaft utilizes three dynamic dampers.
4—Dynamic dampers are designed to resonate at the natural frequency of the crankshaft.

4014. On which stroke or strokes are both valves on a four–stroke cycle reciprocating engine cylinder open?

1—Exhaust.
2—Intake.
3—Power and intake.
4—Exhaust and intake.

4015. When timing a magneto to an engine using a d.c. continuity tester, the primary circuit between the coil and the breaker points should·be opened to prevent the

1—points from welding together.
2—primary coil from burning out if the timing operation is prolonged.
3—permanent magnet from becoming neutralized.
4—condenser action from interfering with the timing operation.

4016. Cam–ground pistons are installed in some aircraft engines to

1—provide a better fit at operating temperatures.
2—cause the master rod piston to wear at the same rate as those installed on the articulating rods.
3—act as a compensating feature so that a compensated magneto is not required.
4—equalize the wear on pistons that do not operate in a vertical plane.

4017. Using the following information, determine how many degrees the crankshaft will rotate with both the intake and exhaust valves seated.

Intake opens 15° B.T.D.C.
Exhaust opens 70° B.B.D.C.
Intake closes 45° A.B.D.C.
Exhaust closes 10° A.T.D.C.

1—610°.
2—290°.
3—245°.
4—25°.

4018. An overhead valve engine using zero–lash hydraulic valve lifters is observed to have no clearance in its valve–operating mechanism after the minimum inlet oil and cylinder head temperatures for takeoff have been reached. When can this condition be expected?

1—During normal operation.
2—When the lifters become deflated.
3—As a result of carbon and sludge becoming trapped in the lifter and restricting its motion.
4—As a result of inverting the tappet valve during assembly of the lifter.

4019. What tool is generally used to measure the crankshaft rotation in degrees?

1—Dial indicator.
2—Top–center indicator.
3—Timing disk.
4—Timing light.

4020. If an engine with a stroke of 6 inches is operated at 2,000 RPM, the piston movement within the cylinder will

1—be at maximum velocity around T.D.C.
2—be constant during the entire 360° of crankshaft travel.
3—be at maximum velocity 90° after T.D.C.
4—average approximately 60 MPH.

4021. The inside of some cylinder barrels is hardened by

1—nitriding.
2—shot peening.
3—nickel plating.
4—cadmium plating.

4022. Which statement is correct regarding a four–stroke cycle aircraft engine?

1—The intake valve closes on the compression stroke.
2—The exhaust valve opens on the exhaust stroke.
3—The intake valve opens on the intake stroke.
4—The exhaust valve closes on the exhaust stroke.

4023. During overhaul, reciprocating engine intake and exhaust valves are checked for stretch

1—with a suitable outside micrometer caliper.
2—with a contour gauge.
3—with a suitable vernier caliper.
4—by placing the valve on a surface plate and measuring its length with a vernier height gauge.

4024. When is the fuel/air mixture ignited in a conventional reciprocating engine?

1—When the piston has reached top dead center of the intake stroke.
2—Just as the piston begins the power stroke.
3—Shortly before the piston reaches the top of the compression stroke.
4—When the piston reaches top dead center compression stroke.

4025. Ignition occurs at 28° B.T.C. on a certain four–stroke cycle engine, and the intake valve opens at 15° B.T.C. How many degrees of crankshaft travel after ignition does the intake valve open? (Consider one cylinder only.)

1—707°.
2—373°.
3—347°.
4—13°.

4026. What is the purpose of the safety circlet generally installed on valve stems?

1—To hold the valve guide in position.
2—To hold the valve spring retaining washer in position.
3—To prevent exhaust gases from entering the rocker box chamber.
4—To prevent valves from falling into the combustion chamber.

4027. When timing the valves of a fully assembled radial engine, what will be the result of failure to eliminate any backlash that may exist in the mechanism?

1—Valve lift will be less than specified.
2—Inaccurate valve timing.
3—Valve lift will be more than specified.
4—Valve lap will be reduced.

4028. The operating valve clearance of an engine using hydraulic tappets (zero lash lifters) should not exceed

1—0.15 to 0.18 inch.
2—0.00 inch.
3—0.25 to 0.32 inch.
4—0.30 to .110 inch.

4029. If the exhaust valve of a four–stroke cycle engine is closed and the intake valve is just closing, the piston is on the

1—intake stroke.
2—power stroke.
3—exhaust stroke.
4—compression stroke.

4030. How many of the following are factors in establishing the maximum compression ratio limitations of an aircraft engine?

(1) Detonation characteristics of the fuel used.
(2) Design limitations of the engine.
(3) Degree of supercharging.
(4) Thermal efficiency of the engine.

1—One.
2—Four.
3—Two.
4—Three.

4031. Full–floating piston pins are those which allow motion between the pin and

1—the piston.
2—both the piston and connecting rod.
3—neither the piston nor the connecting rod.
4—the connecting rod.

4032. For what purpose are the intake and exhaust valves of some engines designed to overlap?

1—To allow the engine to operate at a higher RPM.
2—To allow the use of a four–lobe cam ring.
3—To promote ease of starting.
4—To improve the volumetric efficiency of the engine.

4033. If the hot clearance is used to set the valves when the engine is cold, what will occur during operation of the engine?

1—The valves will open early and close early.
2—The valves will open late and close early.
3—The valves will open early and close late.
4—No ill effects will occur.

4034. The purpose of two or more valve springs in aircraft engines is to

1—reduce valve stretch.
2—equalize side pressure on the valve stems.
3—eliminate valve spring surge.
4—eliminate valve stem breakage.

4035. Top overhaul of a piston engine means

1—complete reconditioning of engine and accessories.
2—ignition tuning and adjustment of valve clearances.
3—reconditioning the cylinders, pistons, and valve–operating mechanism.
4—replacement of cylinder rings and rod bearings.

4036. Why does the smoothness of operation of an engine increase with a greater number of cylinders?

1—The power impulses are spaced closer together.
2—The engine is heavier.
3—The number of cylinders has nothing to do with the smoothness of operation.
4—The engine has larger counter balance weights.

4037. Compression ratio is the ratio between the

1—piston travel on the compression stroke and on the intake stroke.
2—combustion chamber pressure on the combustion stroke and on the exhaust stroke.
3—cylinder volume with piston at bottom dead center and at top dead center.
4—fuel and air in the combustion chamber.

4038. If the crankshaft run–out readings on the dial indicator are plus .002 inch and minus .003 inch, the runout is

1—.005 inch.
2—.001 inch.
3—plus .001 inch.
4—minus .001 inch.

4039. Which of the following should be checked when inspecting engine ball bearings?

1—Proper degree of hardness.
2—Metal dissimilation.
3—Bearing out–of–balance.
4—Flaking or pitting of races.

4040. How is proper end–gap clearance on new piston rings assured during the major overhaul of an engine?

1—By using a go and no–go gauge.
2—By using rings specified by the engine manufacturer.
3—By placing the rings in the cylinder and measuring the end–gap with a feeler gauge.
4—By grinding the rings on an emery wheel.

4041. The volume of a cylinder equals 70 cubic inches when the piston is at bottom center. When the piston is at the top of the cylinder, the volume equals 10 cubic inches. What is the compression ratio?

1—10:7.
2—1:7.
3—7:10.
4—7:1.

4042. What is the purpose of a power check on a reciprocating engine?

1—To check magneto drop.
2—To check the propeller governor.
3—To determine satisfactory performance.
4—To determine if the fuel/air mixture is adequate.

4043. When checking compression with the differential pressure tester, the test cannot be made with the piston at bottom dead center because

1—it is too dangerous.
2—the cylinder volume is at its maximum, thus giving the incorrect reading.
3—you may damage the gauge.
4—at any bottom dead center position at least one valve will be open.

4044. Which of the following will be caused by excessive valve clearance of a cylinder on a reciprocating aircraft engine?

1—Reduced valve overlap period.
2—Increased cylinder pressure on the power stroke.
3—Intake and exhaust valves will open early and close late.
4—A power increase by shortening the exhaust event.

4045. What is the probable cause for oil being thrown out the breather on a wet–sump reciprocating engine?

1—Inoperative scavenger pump.
2—Worn piston rings.
3—Oil pressure relief valve inoperative.
4—Excessive oil quantity.

4046. Which of the following would indicate a general weak–engine condition when operated with a fixed–pitch propeller or test club?

1—Oil pressure lower at idle RPM than at cruise RPM.
2—Lower than normal static RPM, full throttle operation.
3—Manifold pressure lower at idle RPM than at static RPM.
4—Lower than normal manifold pressure for any given RPM.

4047. Which of the following is required by FAR Part 43 when performing a 100–hour inspection on reciprocating engines?

1—Magneto timing check.
2—Cylinder compression check.
3—Valve timing check.
4—Crankshaft run–out check.

4048. One engine of a multiengine aircraft must be shut down because of high operating temperatures, loss of power, loss of oil through the engine breather, and complete loss of oil pressure. What is the most likely cause?

1—An inoperative engine oil supply pump.
2—An inoperative scavenger pump.
3—A ruptured supercharger shaft oil seal.
4—A clogged crankshaft oil passageway.

4049. As the pressure is applied during a reciprocating engine compression check using a differential pressure tester, what would a movement of the propeller in the direction of engine rotation indicate?

1—The piston was positioned ahead of top dead center.
2—The piston was on compression stroke.
3—The piston was on intake stroke.
4—The piston was positioned past top dead center.

4050. Excessive valve clearance results in the valves opening

1—early and closing early.
2—late and closing early.
3—early and closing late.
4—late and closing late.

4051. During routine inspection of a reciprocating engine a deposit of small, bright, metallic particles which do not cling to the magnetic drain plug is discovered in the oil sump and on the surface of the oil filter. This condition

1—may be a result of abnormal plain type bearing wear and is cause for further investigation.
2—indicates accessory section gear wear and is cause for removal and/or overhaul.
3—is probably a result of ring and cylinder wall wear and is cause for engine removal and/or overhaul.
4—is normal in engines utilizing plain type bearings and aluminum pistons and is not cause for alarm.

4052. Select the speed and direction of rotation of a four–lobe cam plate in relation to the crankshaft in a nine–cylinder radial engine.

1—One–eighth crankshaft speed and same direction.
2—One–half crankshaft speed and opposite direction.
3—One–eighth crankshaft speed and opposite direction.
4—One–half crankshaft speed and same direction.

4053. What is the minimum number of crankshaft revolutions required to cause the five–lobe cam plate of a nine–cylinder radial engine to turn one complete revolution?

1—Ten.
2—Two.
3—Four and one–half.
4—Five.

4054. What is the purpose of dynamic suspension as applied to aircraft reciprocating engine installations?

1—To eliminate the torsional flexibility of the powerplant.
2—To reduce the amplitude of the normal engine vibrations.
3—To make the powerplant installation more rigid.
4—To isolate normal powerplant vibrations from the aircraft structure.

4055. If metallic particles are found on the oil screen during an engine inspection,

1—it is an indication of normal engine wear unless the particles are nonferrous.
2—the cause should be identified and corrected before the aircraft is released for flight.
3—it is an indication of normal engine wear unless the deposit exceeds a specified amount.
4—it is an indication of normal engine wear unless the particles show ferritic content (respond to a magnet).

4056. If the oil pressure gauge fluctuates over a wide range from zero to normal operating pressure, the most likely cause is

1—malfunction of the thermostatic control valve.
2—low oil supply.
3—broken or weak pressure relief valve spring.
4—air lock in the scavenge pump intake.

4057. What special procedure must be followed when adjusting the valves of an engine equipped with a floating cam ring?

1—Adjust valves when the engine is hot.
2—Adjust all exhaust valves before intake valves.
3—Eliminate cam bearing clearance when making valve adjustment.
4—Adjust all intake valves before exhaust valves.

4058. Which of the following is most likely to occur if an overhead valve engine is operated with inadequate valve clearances?

1—The valves will not open during start and engine warmup.
2—The valves will remain closed for longer periods than specified by the engine manufacturer.
3—The valves will not seat positively during start and engine warmup.
4—The further decrease in valve clearance that occurs as engine temperatures increase will cause damage to the valve–operating mechanism.

4059. Excessive valve clearances will cause the duration of valve opening to

1—increase for both intake and exhaust valves.
2—decrease for both intake and exhaust valves.
3—decrease for intake valves and increase for exhaust valves.
4—increase for intake valves and decrease for exhaust valves.

4060. What does valve overlap promote?

1—Lower intake manifold pressure and temperatures.
2—A backflow of gases across the cylinder.
3—An overlap of the power and intake strokes.
4—Better scavenging and cooling characteristics.

4061. The indicated oil pressure of a particular dry–sump aircraft engine is higher at cruise RPM than at idle RPM. This indicates

1—defective piston–oil control rings.
2—excessive relief–valve spring tension.
3—an insufficient oil supply.
4—normal operation.

4062. At what speed must a crankshaft turn if each cylinder of a four–stroke cycle engine is to be fired 200 times a minute?

1—200 RPM.
2—800 RPM.
3—1,600 RPM.
4—400 RPM.

4063. Crankshaft run–out is checked

1—after each flight and after a 30–day layoff.
2—during engine overhaul and in case of sudden stoppage of the engine.
3—during engine overhaul and anytime it is convenient.
4—if the propeller is too noisy and vibrates.

4064. Before attempting to start a radial engine that has been shut down for more than 30 minutes,

1—place the fuel selector valve in the OFF position.
2—pull the propeller through by hand in the opposite direction of normal rotation to check for liquid lock.
3—turn the ignition switch on before energizing the starter.
4—turn the propeller three to four revolutions in the normal direction of rotation to check for liquid lock.

4065. Which of the following cam rings will turn the slowest relative to the crankshaft?

1—One–lobe cam ring used on a 14–cylinder engine.
2—Two–lobe cam ring used on a five–cylinder engine.
3—Three–lobe cam ring used on a seven–cylinder engine.
4—Four–lobe cam ring used on a nine–cylinder engine.

4066. An engine misses in both the right and left positions of the magneto switch. The quickest method for locating the trouble is to

1—check for cold cylinders to isolate the trouble.
2—perform a compression check.
3—check for a weak breaker spring in the magneto.
4—check each spark plug.

4067. A hissing sound from the exhaust stacks when the propeller is being pulled through manually indicates

1—a cracked exhaust stack.
2—exhaust valve blow–by.
3—worn piston rings.
4—liquid lock.

4068. If the oil pressure of a cold engine is higher than at normal operating temperatures, the

1—oil system relief valve should be readjusted.
2—engine's lubrication system is probably operating normally.
3—oil dilution system should be turned on immediately.
4—engine should be shut down immediately.

4069. If an engine operates with a low oil pressure and a high oil temperature, the problem may be caused by a

1—low setting of the oil thermostat.
2—leaking oil dilution valve.
3—sheared oil pump shaft.
4—clogged oil cooler annular jacket.

4070. Which fuel/air mixture will result in the highest engine temperature (all other factors remaining constant)?

1—A mixture leaner than a rich best–power mixture of .085.
2—A mixture richer than a lean best–power mixture of .075.
3—A mixture richer than a full–rich mixture of .087.
4—A mixture leaner than a manual lean mixture of .060.

4071. If an engine cylinder is to be removed, at what position in the cylinder should the piston be?

1—Bottom dead center.
2—Top dead center.
3—Halfway between top and bottom dead center.
4—Any convenient position.

4072. The operating valve clearance of a radial engine as compared to cold valve clearance is

1—greater.
2—less.
3—the same.
4—greater or less depending on the type of valve used.

4073. What is the firing order for a nine–cylinder radial engine?

1—1,2,3,4,5,6,7,8,9.
2—1,2,3,8,4,7,5,6,9.
3—1,3,5,7,9,2,4,6,8.
4—9,4,2,7,5,6,3,1,8.

4074. Engine operating flexibility is the ability of the engine to

1—deliver maximum horsepower at a specific altitude.
2—meet exacting requirements of efficiency and low weight per horsepower ratio.
3—run smoothly and give the desired performance at all speeds.
4—expand and contract with changes in temperature and pressure.

4075. Standard aircraft cylinder oversizes usually range from 0.010 inch to 0.030 inch. Oversize on automobile engine cylinders may range up to 0.100 inch. This is because aircraft engine cylinders

1—are limited as to the range of piston sizes available.
2—have relatively thin walls and may be nitrided.
3—cannot have ridging removed by grinding.
4—operate at high temperatures.

4076. If the ignition switch is moved from BOTH to either LEFT or RIGHT during an engine ground check, normal operation is usually indicated by

1—a large drop in RPM.
2—a slight increase in RPM.
3—no change in RPM.
4—a slight drop in RPM.

4077. During ground check an engine is found to be rough-running, the magneto drop is normal, and the manifold pressure is higher than normal for any given RPM. The trouble may be caused by

1—a loose connection on the high-tension lead to one magneto.
2—several spark plugs fouled on different cylinders.
3—a leak in the intake manifold.
4—a dead cylinder.

4078. What is the best indication of worn valve guides?

1—High oil consumption.
2—Low compression.
3—Low oil pressure.
4—High oil pressure.

4079. By use of a differential pressure compression tester, it is determined that the No. 3 cylinder of a nine-cylinder radial engine will not hold pressure after the crankshaft has been rotated 260° from top dead center compression stroke No. 1 cylinder. How can this indication usually be interpreted?

1—Badly worn or damaged piston rings.
2—A normal indication.
3—Exhaust valve blow-by.
4—A damaged exhaust valve or insufficient exhaust valve clearance.

4080. What effect will an increase in manifold pressure with a constant RPM have on the bearing load between the crankshaft and master rod bearing in a single-row radial engine?

1—The load will decrease.
2—The load will remain the same.
3—The effect on the bearing load cannot be determined from the information given.
4—The load will increase.

4081. Direct mechanical push-pull carburetor heat control linkages should normally be adjusted so that the stop located on the diverter valve will be contacted

1—before the stop at the control lever is reached in both HOT and COLD positions.
2—before the stop at the control lever is reached in the HOT position and after the stop at the control lever is reached in the COLD position.
3—after the stop at the control lever is reached in both HOT and COLD positions.
4—after the stop at the control lever is reached in the HOT position and before the stop at the control lever is reached in the COLD position.

4082. Reduced air density at high altitude has a decided effect on carburetion, resulting in a reduction of engine power by

1—excessively enriching the fuel/air mixture.
2—excessively leaning the fuel/air mixture.
3—decreasing the volatility of the fuel.
4—increasing the pressure differential between the carburetor and the intake manifold.

4083. Increased water vapor (higher relative humidity) in the incoming air to a reciprocating engine will normally result in which of the following?

1—Decreased engine power at a constant RPM and manifold pressure.
2—Increased power output due to increased volumetric efficiency.
3—Reduced fuel flow requirements at high-power settings due to reduced detonation tendencies.
4—A leaning effect on engines which use non-automatic carburetors.

4084. How does detonation differ from preignition?

1—Preignition occurs in only a few cylinders at one time.
2—Detonation cannot be detected in an engine as easily as preignition.
3—Preignition will cause a loss of power, but will not damage an engine.
4—Detonation usually occurs in only a few cylinders at one time.

4085. Which of the following engine servicing operations generally requires engine pre-oiling prior to starting the engine?

1—Oil filter change.
2—Engine oil change.
3—Engine installation.
4—Replacement of oil lines.

4086. During the inspection of an engine control system in which push-pull control rods are used, the threaded rod ends should

1—not be adjusted in length for rigging purposes because the rod ends have been properly positioned and staked during manufacture.
2—be checked to determine that they are properly safetied to the push-pull rod with brass or stainless steel safety wire.
3—be checked for thread engagement of at least one and one-half threads but not more than three threads.
4—be checked for the amount of thread engagement by means of the inspection holes provided.

4087. Which of the following conditions would most likely lead to detonation?

1—Improper ignition timing.
2—Improper valve grinding at overhaul.
3—Use of fuel with too high an octane rating.
4—Use of fuel with too low an octane rating.

4088. The manifold pressure of an unsupercharged engine, operated at full throttle at sea level, will be less than sea-level pressure. At altitude, providing the RPM is unchanged, the

1—engine will lose power due to the reduced volume of air drawn into the cylinders.
2—power produced by the engine will remain unchanged.
3—power produced by the engine will increase slightly due to the reduced exhaust back pressure.
4—engine will lose power due to the reduced density of the air drawn into the cylinders.

4089. Which of the following would most likely cause a reciprocating engine to backfire through the induction system at low RPM operation?

1—Idle speed too low.
2—Idle mixture too rich.
3—Clogged derichment valve.
4—Lean mixture.

4090. How may it be determined that a reciprocating engine with a dry sump is pre-oiled sufficiently?

1—Oil will appear on the cylinder interior walls.
2—The engine oil pressure gauge will indicate normal oil pressure.
3—Oil will flow from the engine return line or indicator port.
4—When the quantity of oil specified by the manufacturer has been pumped into the engine.

4091. What is the basic operational sequence for reducing the power output of an engine equipped with a constant-speed propeller?

1—Reduce the RPM, then the manifold pressure.
2—Reduce the RPM, then adjust the propeller control.
3—Reduce the manifold pressure, then retard the throttle to obtain the correct RPM.
4—Reduce the manifold pressure, then the RPM.

4092. Which of the following statements pertaining to fuel/air ratios is true?

1—The mixture ratio which gives the best power is richer than the mixture ratio which gives maximum economy.
2—A lean mixture is faster burning than a normal mixture.
3—A rich mixture is faster burning than a normal mixture.
4—The mixture ratio which gives maximum economy may also be designated as best power mixture.

4093. Backfiring through the carburetor generally results from the use of

1—excessive manifold pressure.
2—an excessively lean mixture.
3—excessively atomized fuel.
4—an excessively rich mixture.

4094. What will cause an engine to have an increased tendency to detonate?

1—Using fuels with high combustion rate characteristics.
2—Retarding the spark advance.
3—Decreasing the density of the charge delivered to the cylinders.
4—Using a rich mixture.

4095. When will small induction system air leaks have the most noticeable effect on engine operation?

1—At medium to high cruise power settings.
2—At high RPM.
3—At maximum continuous and takeoff power settings.
4—At low RPM.

4096. To reduce the power output of an engine equipped with a constant-speed propeller and operating near maximum BMEP, the

1—manifold pressure is reduced with the throttle control before the RPM is reduced with the propeller control.
2—manifold pressure is reduced with the propeller control before the RPM is reduced with the throttle control.
3—RPM is reduced with the throttle control before the manifold pressure is reduced with the propeller control.
4—RPM is reduced with the propeller control before the manifold pressure is reduced with the throttle control.

4097. One of the best indicators of reciprocating engine combustion chamber problems is

1—excessive engine vibration.
2—low oil pressure.
3—carburetor condition.
4—spark plug condition.

4098. What could cause excessive pressure buildup in the crankcase of a reciprocating engine?

1—Plugged crankcase breather.
2—Oil pump pressure adjusted too high.
3—An excessive quantity of oil.
4—Worn oil scavenger pump.

4099. Excessive valve clearance in a piston engine

1—increases valve overlap.
2—has no effect on valve overlap.
3—increases valve service life.
4—decreases valve overlap.

4100. The critical altitude is the highest altitude at which an engine will maintain, at the maximum continuous rotational speed, maximum

1—peak horsepower.
2—brake horsepower.
3—continuous horsepower.
4—cruise horsepower.

4101. At what point in an axial–flow turbojet engine will the highest gas pressures occur?

1—Immediately after the turbine section.
2—At the turbine entrance.
3—Within the burner section.
4—At the compressor outlet.

4102. Identify a function of the nozzle diaphragm in a turbojet engine.

1—To decrease the velocity of exhaust gases.
2—To center the fuel spray in the combustion chamber.
3—To direct the flow of gases to strike the turbine buckets at a desired angle.
4—To direct the flow of gases into the combustion chamber.

4103. What is the profile of a turbine engine compressor blade?

1—The shape of the blade root at the disk attachment.
2—The leading edge of the blade.
3—A cutout that reduces blade tip thickness.
4—The curvature of the blade root.

4104. The fan rotational speed of a dual axial compressor forward fan engine is the same as the

1—accessory drive shaft.
2—low–pressure compressor.
3—forward turbine wheel.
4—high–pressure compressor.

4105. The abbreviation "P" with subscript t7 used in jet engine terminology means

1—the total inlet pressure.
2—pressure and temperature at station No. 7.
3—seven times the temperature divided by the total pressure.
4—the total presssure at station No. 7.

4106. What is the function of the nozzle diaphragm (gas turbine engine) located on the upstream side of the turbine wheel?

1—To increase the pressure of the exhaust mass.
2—To increase the velocity of the heated gases flowing past the nozzle diaphragm.
3—To direct the flow of gases parallel to the chord line of the turbine buckets.
4—To decrease the velocity of the heated gases flowing past the nozzle diaphragm.

4107. What turbine engine section provides for proper mixing of the fuel and air?

1—Combustion section.
2—Compressor section.
3—Turbine section.
4—Accessory section.

4108. In a gas turbine engine, combustion occurs at a constant

1—volume.
2—pressure.
3—velocity.
4—density.

4109. Which of the following statements is true regarding gas turbine engines?

1—At the lower engine speeds, thrust increases rapidly with small increases in RPM.
2—At the higher engine speeds, thrust increases rapidly with small increases in RPM.
3—Gas turbine engines operate less efficiently at high altitudes due to the lower temperature encountered.
4—The thrust delivered per pound of air consumed is less at high altitude than at low altitude.

4110. Some high–volume turboprop and turbojet engines are equipped with two–spool or split compressors. When these engines are operated at high altitudes, the

1—throttle must be retarded to prevent overspeeding of the two compressor rotors due to the lower density air.
2—low–pressure rotor will increase in speed as the compressor load decreases in the lower density air.
3—throttle must be retarded to prevent overspeeding of the high–pressure rotor due to the lower density air.
4—low–pressure rotor will decrease in speed as the compressor load decreases in the lower density air.

4111. Gas turbine engines use a nozzle diaphragm which is located on the upstream side of the turbine wheel. One of the functions of this unit is to

1—decrease the velocity of the heated gases flowing past this point.
2—direct the flow of gases parallel to the vertical line of the turbine buckets.
3—increase the velocity of the heated gases flowing past this point.
4—increase the pressure of the exhaust mass.

4112. Where is the highest gas pressure in a turbojet engine?

1—At the outlet of the tailpipe section.
2—At the entrance of the turbine section.
3—In the entrance of the burner section.
4—In the outlet of the burner section.

4113. An exhaust cone placed aft of the turbine in a jet engine will cause the pressure to

1—increase and the velocity to decrease.
2—increase and the velocity to increase.
3—decrease and the velocity to increase.
4—decrease and the velocity to decrease.

4114. What is the function of the stator vane assembly at the discharge end of a typical axial–flow compressor?

1—To reduce drag on the first stage turbine blades.
2—To straighten airflow to eliminate turbulence.
3—To direct the flow of gases into the combustion chambers.
4—To increase air swirling motion into the combustion chambers.

4115. The turbines near the rear of a jet engine

1—compress air heated in the combustion section.
2—increase air velocity for propulsion.
3—circulate air to cool the engine.
4—drive the compressor section.

4116. When starting a turbojet engine,

1—a hot start is indicated if the exhaust gas temperature exceeds specified limits.
2—an excessively lean mixture is likely to cause a hot start.
3—the engine should start between 60 to 80 seconds after the fuel shutoff lever is opened.
4—release the starter switch as soon as indication of light–off occurs.

4117. In the dual axial–flow or twin spool compressor system, the first stage turbine drives the

1—N_1 and N_2 compressors.
2—N_4 compressor.
3—N_2 compressor.
4—N_1 compressor.

4118. Cracks may occur in hot section components of a turbine engine if they are marked during inspection with

1—a lead pencil.
2—chalk.
3—layout dye.
4—any of the above.

4119. What drives the fan in most turbofan engines?

1—The turbine that drives the high pressure compressor.
2—The turbine that drives the low pressure compressor.
3—A special turbine that drives nothing but the fan.
4—An electric motor driven by electricity generated by the starter–generator.

4120. When starting a turbine engine, a hung start is indicated if the engine

1—exhaust gas temperature exceeds specified limits.
2—fails to reach idle RPM.
3—RPM exceeds specified operating speed.
4—pressure ratio exceeds specified operating limits.

4121. What are the two main sections of a turbine engine for inspection purposes?

1—Combustion and exhaust.
2—Hot and cold.
3—Compressor and turbine.
4—Combustion and turbine.

4122. What are the two basic elements of the turbine section in a turbine engine?

1—Impeller and diffuser.
2—Compressor and manifold.
3—Bucket and expander.
4—Stator and rotor.

4123. What is the primary function of the exhaust cone of a turbine engine?

1—Collect and convert exhaust gases into a solid low velocity exhaust vapor.
2—Straighten the swirling exhaust gases.
3—Collect and convert exhaust gases into a solid high velocity exhaust jet.
4—Pipe the exhaust gases out of the airframe.

4124. What are the two functional elements in a centrifugal compressor?

1—Turbine and compressor.
2—Compressor and manifold.
3—Bucket and expander.
4—Impeller and diffuser.

4125. What must be done after the fuel control unit has been replaced on a turbine engine?

1—Retime the engine.
2—Recalibrate the fuel nozzles.
3—Retrim the engine.
4—Recheck the flame pattern.

4126. What is the most satisfactory method of attaching turbine blades to turbine wheels?

1—The fir–tree design.
2—The tongue and groove design.
3—High temp–high strength adhesive method.
4—Press fit method.

4127. A turbine engine compressor which contains vanes on both sides of the impeller is a

1—single entry centrifugal compressor.
2—double entry centrifugal compressor.
3—double entry axial–flow compressor.
4—single entry axial–flow compressor.

4128. What is the first engine instrument indication of a successful start of a turbine engine?

1—A decrease in the exhaust gas temperature.
2—A rise in the engine fuel flow.
3—A decrease in the engine pressure ratio.
4—A rise in the exhaust gas temperature.

4129. How does a dual axial–flow compressor improve the efficiency of a turbojet engine?

1—More turbine wheels can be used.
2—Combustion chamber temperatures are reduced.
3—Higher compression ratios can be obtained.
4—The velocity of the air entering the combustion chamber is increased.

4130. Two basic types of turbine blades are

1—reaction and converging.
2—tangential and reaction.
3—reaction and impulse.
4—impulse and vector.

4131. A turboprop powerplant propeller

1—is governed at the same speed as the turbine.
2—controls the speed of the engine in the beta range.
3—accounts for 75 to 85 percent of the total thrust output.
4—accounts for 15 to 25 percent of the total thrust output.

4132. An advantage of the axial–flow compressor is its

1—low starting power requirements.
2—low weight.
3—high peak efficiency.
4—high frontal area.

4133. What is the purpose of the stator blades in the compressor section of a turbine engine?

1—Stabilize pressure.
2—Prevent compressor surge.
3—Control direction of the airflow.
4—Increase velocity of the airflow.

4134. What is the purpose of the diffuser section in a turbine engine?

1—To increase pressure and reduce velocity.
2—To speed up the airflow in the turbine section.
3—To convert pressure to velocity.
4—To reduce pressure and increase velocity.

4135. Where do stress rupture cracks usually appear on turbine blades of turbojet engines?

1—Across the blade root, parallel to the fir tree.
2—Along the trailing edge, parallel to the edge.
3—Along the leading edge, parallel to the edge.
4—Across the leading or trailing edge at a right angle to the edge length.

4136. In which type of turbine engine combustion chamber is the case and liner removed and installed as one unit during routine maintenance?

1—Can.
2—Can annular.
3—Variable.
4—Annular.

4137. The diffuser section of a jet aircraft engine is located between

1—the burner section and the turbine section.
2—the N_1 section and the N_2 section.
3—station No. 7 and station No. 8.
4—the compressor section and the burner section.

4138. Which of the following are the most common types of thrust reversers used on turbine–engine–powered aircraft?

1—Convergent and divergent.
2—Rotary air vane and stationary air vane.
3—Mechanical blockage and aerodynamic blockage.
4—Cascade vane and blocked door.

4139. When the leading edge of a first–stage turbine blade is found to have stress rupture cracks, which of the following should be suspected?

1—Airseal wear.
2—Faulty cooling shield.
3—Overtemperature condition.
4—Overspeed condition.

4140. Damage to turbine vanes is apt to be greater than damage to compressor vanes because turbine vanes are subjected to much greater

1—stress in the combustor.
2—heat stress.
3—thrust clearance.
4—vibration and other stresses.

4141. Which of the following is the ultimate limiting factor of turbojet engine operation?

1—Compressor inlet air temperature.
2—Compressor outlet air temperature.
3—Turbine inlet temperature.
4—Burner–can pressure.

4142. How is the turbine shaft usually joined to the compressor rotor of a centrifugal compressor turbine engine?

1—Bolted coupling.
2—Keyed coupling.
3—Welded coupling.
4—Splined coupling.

4143. Which of the following engine variables is the most critical during turbine engine operation?

1—Compressor inlet air temperature.
2—Compressor RPM.
3—Burner–can pressure.
4—Turbine inlet temperature.

4144. Reduced blade vibration and improved airflow characteristics in gas turbines are brought about by

1—fir tree blade attachment.
2—impulse type blades.
3—shrouded turbine rotor blades.
4—bulb root attachment.

4145. Which of the following turbojet engine compressors offers the greatest advantages for both starting flexibility and improved high–altitude performance?

1—Single–stage, centrifugal–flow.
2—Dual–stage, centrifugal–flow.
3—Split–spool, axial–flow.
4—Single–spool, axial–flow.

4146. Turbojet engine turbine blades removed for detailed inspection must be re–installed in

1—a slot 180° away.
2—a slot 90° clockwise.
3—a slot 90° counterclockwise.
4—the same slot.

4147. An advantage of the centrifugal–flow compressor is its high

1—frontal area.
2—pressure rise per stage.
3—ram efficiency.
4—peak efficiency.

4148. The highest heat–to–metal contact in a jet engine is the

1—burner cans.
2—exhaust cone.
3—turbine inlet guide vanes.
4—turbine blades.

4149. Which of the following two elements make up the axial–flow compressor assembly?

1—Rotor and stator.
2—Rotor and diffuser.
3—Compressor and manifold.
4—Stator and diffuser.

4150. The two types of centrifugal compressor impellers are

1—single stage and two stage.
2—single entry and double entry.
3—rotor and stator.
4—impeller and diffuser.

4151. Between each row of rotating blades in a turbine engine compressor, there is a row of stationary blades which act to diffuse the air. These stationary blades are called

1—buckets.
2—expanders.
3—diffuser blades.
4—stators.

4152. Standard sea level pressure is

1—30.92 inches of mercury.
2—32.174 inches of mercury.
3—56.2 inches of mercury.
4—29.92 inches of mercury.

4153. Using standard atmospheric conditions, the standard sea level temperature is

1—40° F.
2—0° F.
3—0° C.
4—15° C.

4154. When aircraft turbine blades are subjected to excessive temperatures, what type of failures would you expect?

1—Compression and torsion.
2—Bending and torsion.
3—Torsion and tension.
4—Stress rupture.

4155. In an axial–flow compressor, one purpose of the stator vanes at the discharge end of the compressor is to

1—prevent compressor surge and eliminate stalls.
2—straighten the airflow and eliminate turbulence.
3—increase the velocity and prevent swirling and eddying.
4—decrease the velocity, prevent swirling, and decrease pressure.

4156. Dirty compressor blades in a turbine engine could result in

1—low RPM.
2—low EGT.
3—high RPM.
4—high EGT.

4157. The two types of compressors most commonly used in jet engines are

1—axial and root.
2—centrifugal and reciprocating.
3—root and centrifugal.
4—centrifugal and axial.

4158. A purpose of the shrouds on the turbine blades of an axial–flow engine is to

1—reduce vibration.
2—shorten run–in time.
3—increase tip speed.
4—reduce air entrance.

4159. In a dual axial–flow compressor, the first stage turbine drives

1—N_2 compressor.
2—N_1 compressor.
3—low pressure compressor.
4—both low and high pressure compressors.

4160. What should be done if a turbine engine catches fire during starting?

1—Turn off the fuel and continue cranking.
2—Disengage starter immediately.
3—Continue starting attempt to blow out fire.
4—Place power lever in increase to exhaust fuel fumes.

4161. What is the proper starting sequence for a turbojet engine?

1—Ignition, starter, fuel.
2—Fuel, starter, ignition.
3—Starter, ignition, fuel.
4—Starter, fuel, ignition.

4162. Inflight turbine engine flameouts are usually caused by

1—high exhaust gas temperature.
2—interruption of the inlet airflow.
3—fouling of the primary igniter plugs.
4—fuel–nozzle clogging.

4163. What units in a gas turbine engine aid in stabilization of the compressor during low thrust engine operations?

1—Bleed air valves.
2—Stator vanes.
3—Inlet guide vanes.
4—Pressurization and dump valves.

4164. In a turbine engine with a dual–axial compressor, the low speed compressor

1—always turns at the same speed as the high speed compressor.
2—is connected directly to the high speed compressor.
3—seeks its own best operating speed.
4—has a higher compressor shaft speed than the high speed compressor.

4165. What is the function of the inlet guide vane assembly on a centrifugal compressor?

1—Directs the air into the first stage rotor blades at the proper angle.
2—Converts velocity energy into pressure energy.
3—Converts pressure energy into velocity energy.
4—Picks up air and adds energy as it accelerates outward by centrifugal force.

4166. Hot spots on the tail cone of a turbine engine are possible indicators of a malfunctioning fuel nozzle or a

1—faulty combustion chamber.
2—loose inlet air guide vane.
3—faulty igniter plug.
4—improperly positioned tail cone.

4167. The stator vanes in an axial–flow compressor

1—convert velocity energy into pressure energy.
2—convert pressure energy into velocity energy.
3—direct air into the first stage rotor vanes at the proper angle.
4—pick up air and add energy as it accelerates outward by centrifugal force.

4168. What happens to velocity as air flows through a convergent nozzle?

1—Decreases.
2—Remains constant.
3—Is inversely proportional to the temperature.
4—Increases.

4169. What happens to velocity as air flows through a divergent nozzle?

1—Increases.
2—Remains constant.
3—Is inversely proportional to the temperature.
4—Decreases.

4170. What happens to pressure as air flows through a convergent nozzle?

1—Increases.
2—Decreases.
3—Remains constant.
4—Is inversely proportional to the temperature.

4171. What happens to pressure as air flows through a divergent nozzle?

1—Decreases.
2—Remains constant.
3—Increases.
4—Is inversely proportional to the temperature.

4172. Anti–icing of turbojet engine air inlets is accomplished by

1—electrical heating elements inside the inlet guide vanes.
2—hot air ducted over the outside of the inlet guide vanes.
3—engine bleed air ducted through the critical areas.
4—electrical heating elements located within the engine air inlet cowling.

4173. When starting a turbojet engine, the starter should be disengaged when the

1—engine lights are off.
2—engine reaches idle RPM.
3—RPM indicator shows 100 percent.
4—ignition and fuel system are activated.

4174. What is the primary advantage of an axial–flow compressor over a centrifugal compressor?

1—Easier maintenance.
2—High frontal area.
3—Less expensive.
4—Greater pressure ratio.

4175. What is the purpose of blow–in doors in the induction system of a turbine engine aircraft?

1—Admit air to the engine compartment during ground operation when the engine air requirements are in excess of the amount the normal intake system can supply.
2—Fire extinguisher access openings.
3—Admit air to the engine compartment during flight when the aircraft attitude is not conducive for ram air effect.
4—Access openings for inspection of compressor and turbine blades.

4176. What is meant by a double entry centrifugal compressor?

1—A compressor that has two intakes.
2—A two stage compressor independently connected to the main shaft.
3—Two compressors and two impellers.
4—A compressor with vanes on both sides of the impeller.

4177. What is the major function of the turbine assembly in a turbojet engine?

1—Compresses the air before it enters the combustion section.
2—Directs the gases in the proper direction to the tailpipe.
3—Supplies the power to turn the compressor.
4—Increases the temperature of the exhaust gases.

4178. Stator blades in the compressor section of an axial–flow turbine engine

1—increase the air velocity and prevent swirling.
2—straighten the airflow and accelerate it.
3—decrease the air velocity and prevent swirling.
4—prevent compressor surge.

4179. A gas turbine engine comprises three main sections.

1—Compressor, diffuser, and scavenge.
2—Turbine, combustion, and scavenge.
3—Combustion, compressor, and inlet guide vane.
4—Compressor, combustion, and turbine.

4180. What type of turbine blade is most commonly used in aircraft jet engines?

1—Reaction.
2—Divergent.
3—Impulse.
4—Reaction–impulse.

4181. What is the primary factor which controls the pressure ratio of an axial–flow compressor?

1—Number of stages in compressor.
2—Rotor diameter.
3—Compressor inlet pressure.
4—Compressor inlet temperature.

4182. What are the main sections of a turbojet engine?

1—Fan, combustion, and exhaust.
2—Compressor, combustion, and diffuser.
3—Compressor, combustion, and turbine.
4—Inlet, combustion, and turbine.

4183. What is the possible cause when a turbojet engine indicates no change in power setting parameters, but oil temperature is high?

1—Unusual scavenge pump oil flow.
2—Engine main bearing distress.
3—Gearbox seal leakage.
4—High oil sump pressure.

4184. Which of the following is not a factor in the operation of an automatic fuel control unit used on turbojet engines?

1—Compressor inlet air density.
2—Compressor RPM.
3—Mixture control position.
4—Throttle position.

4185. Newton's First Law of Motion, generally termed the Law of Inertia, states:

1—To every action there is an equal and opposite reaction.
2—Force is proportional to the product of mass and acceleration.
3—Every body persists in its state of rest, or of motion in a straight line, unless acted upon by some outside force.
4—Force applied to an object at any point is transmitted in every direction without loss.

4186. A turbine engine hot section is particularly susceptible to which of the following kind of damage?

1—Scoring.
2—Pitting.
3—Cracking.
4—Galling.

4187. Dirt particles in the air being introduced into the compressor of a turbine engine will form a coating on all but which of the following?

1—Turbine blades.
2—Compressor blades.
3—Casings.
4—Inlet guide vanes.

4188. Severe rubbing of turbine engine compressor blades will usually cause

1—bowing.
2—cracking.
3—burning.
4—galling.

4189. Which of the following influences the operation of an automatic fuel control unit on a turbojet engine?

1—Fuel temperature.
2—Burner pressure.
3—Mixture control position.
4—Exhaust gas temperature.

4190. A turbojet engine having high exhaust gas temperature at desired engine pressure ratio for takeoff indicates

1—that the engine is out of trim.
2—that the fuel control should be replaced.
3—compressor bleed valve malfunction.
4—drain valve malfunction.

4191. The Brayton cycle is known as the constant

1—pressure cycle.
2—volume cycle.
3—temperature cycle.
4—mass cycle.

4192. Where is water injected into a turbojet engine for cooling purposes?

1—Compressor air inlet or diffuser.
2—Second-stage compressor or turbine.
3—Burner can.
4—Fuel control.

4193. Continued and/or excessive heat and centrifugal force on turbine engine compressor blades usually cause

1—profile.
2—growth.
3—gouging.
4—galling.

4194. If the RPM of an axial–flow compressor remains constant, the angle of attack of the blades can be changed by

1—changing the velocity of the airflow.
2—changing the compressor diameter.
3—increasing the pressure ratio.
4—decreasing the pressure ratio.

4195. The compression ratio of an axial–flow compressor is a function of the

1—number of compressor stages.
2—rotor diameter.
3—diffuser area.
4—air inlet velocity.

4196. Which of the following variables affect the inlet air density of a turbine engine?

A. The speed of the aircraft.
B. Compression ratio.
C. Turbine inlet temperature.
D. The altitude of the aircraft.
E. The ambient temperature.
F. The turbine and compressor efficiency.

1—A, C, F.
2—A, D, E.
3—D, E, F.
4—B, C, D.

4197. Which of the following factors affect the thermal efficiency of a turbine engine?

A. Turbine inlet temperature.
B. Compression ratio.
C. The ambient temperature.
D. The speed of the aircraft.
E. Turbine and compressor efficiency.
F. The altitude of the aircraft.

1—C, D, F.
2—E, C, D.
3—A, B, E.
4—A, B, F.

4198. Why do some turbine engines have more than one turbine wheel attached to a single shaft?

1—To facilitate balancing of the turbine assembly.
2—To straighten the airflow before it enters the exhaust area.
3—To help stabilize the pressure between the compressor and the turbine.
4—To extract more power from the exhaust gases than a single wheel can absorb.

4199. The exhaust section of a turbine engine is designed to

1—impart a high exit velocity to the exhaust gases.
2—swirl the exhaust gases.
3—increase temperature, therefore increasing velocity.
4—decrease temperature, therefore decreasing pressure.

4200. Which of the following types of combustion sections are used in aircraft turbine engines?

1—Variable, can–annular, and cascade vane.
2—Annular, variable, and cascade vane.
3—Can, multiple–can, and variable.
4—Multiple–can, annular, and can–annular.

4201. Why does a turbine engine require a cool–off period before shutting it down?

1—To allow the surfaces contacted by the lubricating oil to return to normal operating temperature.
2—To burn off excess fuel ahead of the fuel control.
3—To allow the turbine wheel to cool before the case contracts around it.
4—To avoid seizure of the engine bearings.

4202. How many igniters are normally used on a turbine engine having nine burner cans?

1—One.
2—Two.
3—Three.
4—Nine.

4203. What is meant by a shrouded turbine?

1—The turbine blades are shaped so that their ends form a band or shroud.
2—Each turbine wheel is enclosed by a separate housing or shroud.
3—The turbine wheel is enclosed by a protective shroud to contain the blades in case of failure.
4—The turbine wheel has a shroud or duct which provides cooling air to the turbine blades.

4204. What term is used to describe a permanent and cummulative deformation of the turbine blades of a turbojet engine?

1—Stretch.
2—Elongation.
3—Distortion.
4—Creep.

4205. What is the purpose of the pressurization and dump valve used on turbojet engines?

1—Controls the pressure of the compressor outlet by dumping air when pressure reaches an established level.
2—Allows fuel pressurization of the engine when starting and operating and dumps fuel pressure at engine shutdown.
3—Controls compressor stall by dumping compressor air under certain conditions.
4—Maintains fuel pressure to the fuel control valve and dumps excessive fuel back to the fuel tanks.

4206. At what stage in a turbojet engine are pressures the greatest?

1—Compressor inlet.
2—Turbine outlet.
3—Compressor outlet.
4—Tailpipe.

4207. In what section of a turbojet engine is the jet nozzle located?

1—Combustion.
2—Turbine.
3—Compressor.
4—Exhaust.

4208. Determine which portion of the AD is applicable for Model 0–690 series engine, serial No. 5863–40 with 283 hours' time in service.

This is the compliance portion of an FAA Airworthiness Directive.

Compliance required as indicated:

(A) For Model 0–690 series engines, serial Nos. 101–40 through 5264–40 and IO–690 series engines, serial Nos. 101–48 through 423–48, compliance with (c) required within 25 hours' time in service after the effective date of this AD and every 100 hours' time in service thereafter.

(B) For Model 0–690 series engines, serial Nos. 5265–40 through 6129–40 and IO–690 series engines, serial Nos. 424–48 through 551–48, compliance with (c) required as follows:

(1) Within 25 hours' time in service after the effective date of this AD and every 100 hours' time in service thereafter for engines with more than 275 hours' time in service on the effective date of this AD.

(2) Prior to the accumulation of 300 hours' total time in service and every 100 hours' time in service thereafter for engines with 275 hours' or less time in service on the effective date of this AD.

(C) Inspect the oil pump drive shaft (P/N 67512) on applicable engines in accordance with instructions contained in Connin Service Bulletin No. 295. Any shafts which are found to be damaged shall be replaced before further flight. These inspections shall be continued until Connin P/N 67512 (redesigned) or P/N 74641 oil pump drive shaft is installed at which time the inspections may be discontinued.

1—(B), (1).
2—(A).
3—(B), (2).
4—(A), (B), (C).

4209. A Cessna 180 aircraft has a McCauley propeller Model No. 2A34C50/90A. The propeller is severely damaged in a ground accident, and this model propeller is not available for replacement. Which of the following should be used to find an approved alternate replacement?

1—Summary of Supplemental Type Certificates.
2—Approved aircraft equipment list.
3—Aircraft Specifications/Type Certificate Data Sheets.
4—Aircraft Engine and Propeller Specifications/Type Certificate Data Sheets.

4210. Which of the following is used to monitor the mechanical integrity of the turbines, as well as to check engine operating conditions of a turbojet engine?

1—Engine oil pressure.
2—Exhaust gas temperature.
3—Engine oil temperature.
4—Engine pressure ratio.

4211. The exhaust system on aircraft using a jacket around the engine exhaust as a source of heat should be

1—visually inspected frequently and operational carbon monoxide detection tests performed periodically.
2—replaced at each engine overhaul.
3—replaced at each 100–hour inspection.
4—removed periodically and checked by magnetic–particle inspection.

4212. (1) Airworthiness Directives are Federal Aviation Regulations and must be complied with unless specific exemption is granted.

(2) Airworthiness Directives of an emergency nature require immediate compliance upon receipt.

Regarding the above statements, which of the following is true?

1—Only No. 1 is true.
2—Only No. 2 is true.
3—Neither No. 1 nor No. 2 is true.
4—Both No. 1 and No. 2 are true.

4213. Which of the following contains a minimum checklist for a 100–hour inspection of an engine?

1—FAR Part 91.
2—FAR Part 65.
3—FAR Part 43.
4—Engine Specifications and data sheets.

4214. When must an engine AD (Airworthiness Directive) be complied with after it becomes effective?

1—As specified in the AD.
2—During the next scheduled inspection.
3—At the next scheduled overhaul.
4—Within 30 calendar days.

4215. Which of the following contains a table that lists the engines to which a given propeller is adaptable?

1—Aircraft Type Certificate Data Sheets.
2—Technical Standard Order Authorization.
3—Propeller Type Certificate Data Sheets.
4—Engine Type Certificate Data Sheets.

4216. Which of the following component inspections is to be accomplished on a 100–hour inspection?

1—Check internal timing of magneto.
2—Check cylinder compression.
3—Check float level.
4—Check valve timing.

4217. You are performing a 100–hour inspection on an R985–22 aircraft engine. What does the "985" indicate?

1—The total piston displacement of the engine.
2—The pistons will pump a maximum of 985 cubic inches of air per crankshaft revolution.
3—The total volume of the pistons is 985 plus 22 cubic inches.
4—The total piston displacement of one cylinder.

4218. Agencies whose cylinder barrel chrome plating processes currently are approved by the FAA are normally listed in the

1—Summary of Supplemental Type Certificates.
2—Summary of Airworthiness Directives.
3—Consolidated Listings of FAA Certificated Repair Stations.
4—Aircraft Engine and Propeller Specification and Type Certificate Data Sheets.

4219. During a 100–hour inspection of an R1830–92 engine installed on a DC–3, a mechanic finds "CAA Spec 5E4" stamped on the data plate. Where would the meaning of this stamp be found?

1—Aircraft Specification and Type Certificate Data Sheet.
2—Aircraft Listings.
3—Aircraft Engine Specifications.
4—Aircraft Engine Type Certificate Handbook.

4220. Straightening nitrided crankshafts is

1—recommended.
2—not recommended.
3—approved by repair stations.
4—approved by the manufacturer.

4221. The breaking loose of small pieces of metal from coated surfaces, usually caused by defective plating or excessive loads, is called

1—flaking.
2—chafing.
3—brinelling.
4—pitting.

4222. Each powerplant installed on an airplane with a standard airworthiness certificate must have been

1—installed by the manufacturer.
2—type certificated.
3—manufactured under the TSO system.
4—originally certificated for that aircraft.

4223. A severe condition of chafing or fretting in which a transfer of metal from one part to another occurs, is called

1—gouging.
2—burning.
3—erosion.
4—galling.

4224. Indentations on bearing races caused by high static loads are known as

1—fretting.
2—brinelling.
3—galling.
4—flaking.

4225. What document would be used to determine if a particular engine conforms to its original type design?

1—Federal Aviation Regulations.
2—Engine Manufacturer's Maintenance Manual.
3—Aircraft Engine Specifications or Type Certificata Data Sheets.
4—Checklist in FAR Part 43, Appendix D.

4226. Which of the following can inspect and approve an engine major repair for return to service?

1—Certificated mechanic with airframe and powerplant ratings.
2—Certificated mechanic with a powerplant rating.
3—Certificated mechanic with inspection authorization.
4—Designated Mechanic Examiner.

4227. What publication would be used for guidance to determine whether a powerplant repair is major or minor?

1—Airworthiness Directives.
2—Federal Aviation Regulations, Part 43, Appendix A.
3—Supplemental Type Certificates.
4—Technical Standard Orders.

4228. The airworthiness standards for the issue of type certificates for small airplanes with less than 10 passenger seats in the normal, utility, and acrobatic categories may be found in the

1—Technical Standard Orders (TSO's).
2—Federal Aviation Regulations, Part 23.
3—FAA Advisory Circulars.
4—Minimum Equipment List (MEL) for the specific aircraft.

4229. Which of the following would contain approved data for performing a major repair to an aircraft engine?

1—Engine Type Certificate Data Sheets.
2—Supplemental Type Certificates.
3—Technical Standard Orders.
4—Manufacturer's Mantenance Manual when FAA approved.

4230. What maintenance record(s) are required following a major repair of an aircraft engine?

1—Entries in the airplane flight manual and aircraft logbook.
2—Entries in engine maintenance records and a list of discrepancies for the FAA.
3—Entries in the engine maintenance record and FAA Form 337.
4—Entry in logbook.

4231. A ground incident that results in propeller sudden stoppage may require a crankshaft run–out inspection. What publication would be used to obtain crankshaft run–out tolerance?

1—Federal Aviation Regulations.
2—Current Manufacturer's Maintenance Manual or Instructions for Continued Airworthiness.
3—Type Certificate Data Sheet.
4—AC 43.13–1A Acceptable Methods, Techniques and Practices, Aircraft Inspection and Repair.

4232. Select the Airworthiness Directive applicability statement which applies to an IVO–355 engine, serial number T8164, with 2,100 hours total time and 300 hours since rebuilding.

1—Applies to all IVO–355 engines, serial numbers T8000 through T8300, having less than 2,400 hours total time.
2—Applies to all IVO–355 engines, SN T8000 through T8900 with 2,400 hours or more total time.
3—Applies to all I.O. and TV10–355 engines, all serial numbers regardless of total time or since overhaul.
4—Applies to all IVO–355 engines, serial numbers T4000 through T7999 having more than 2,400 hours total time.

4233. What publication would contain time or cycle limitations for components or parts of a turbine engine installed on a specific aircraft?

1—Instructions for continued airworthiness issued by the airplane manufacturer.
2—Federal Aviation Regulation, Part 33, Airworthiness Standards; Aircraft Engines.
3—Engine Manufacturer Parts Catalog.
4—FAA Advisory Circular 43.13–1A.

4234. How are discharge nozzles in a fuel injected reciprocating engine identified to indicate the flow range?

1—By an identification letter stamped on one of the hexes of the nozzle body.
2—By drilled radial holes connecting the upper counterbore with the outside of the nozzle body.
3—By an identification metal tag attached to the nozzle body.
4—By color codes on the nozzle body.

4235. Which unit most accurately indicates fuel consumption of a reciprocating engine?

1—Fuel flowmeter.
2—BMEP indicator.
3—Fuel pressure gauge.
4—Electronic fuel quantity indicator.

4236. Where is the fuel–flow transmitter usually located in a typical engine fuel system?

1—In the fuel line between the fuel tank shutoff valve and the fuel pump.
2—In the fuel line between the electrical fuel pump and the engine–driven pump.
3—In the fuel line between the fuel tank and the fuel crossfeed valve.
4—In the fuel line between the engine fuel pump and the carburetor.

4237. The type of instrument indicating system that is used in a fuel–flow system in large airplanes is usually

1—a direct reading system.
2—an autosyn system.
3—a synchro resolver system.
4—a servomechanism system.

4238. The current required to operate an aircraft autosyn fuel–flow indicating system is

1—direct current.
2—pulsating current.
3—alternating current.
4—pulsating voltage.

4239. Fuel–flow transmitters are designed to transmit data

1—mechanically.
2—electrically.
3—visually.
4—fluidly.

4240. The fuel–flow transmitter converts the fuel flow into an electrical signal which represents the rate of fuel flow in pounds per hour. It then transmits the signal to

1—a receiver device on the instrument panel.
2—the fuel quantity gauges.
3—the control valve on the fuel control regulator.
4—the bypass valve solenoid on the fuel control regulator.

4241. Which unit most accurately indicates fuel consumption of an internal combustion engine?

1—Fuel quantity gauge.
2—Fuel totalizer.
3—Electronic fuel quantity indicator.
4—Fuel flowmeter.

4242. If a ratiometer–type oil temperature indicator moves off–scale on the high side of the dial as soon as the master switch is turned on, what is the most probable trouble?

1—A short in the power circuit.
2—An open in the bulb circuit.
3—An open in the power circuit.
4—A short in the bulb circuit.

4243. The fuel–flow indicator needle is driven by

1—a friction clutch on the motor shaft.
2—a mechanical gear train.
3—direct coupling to the motor shaft.
4—a magnetic linkage.

4244. On a twin–engine aircraft with fuel injected reciprocating engines, one fuel–flow indicator reads considerably higher than the other in all engine operating configurations. What is the probable cause of this indication?

1—Carburetor icing.
2—One or more fuel nozzle(s) are clogged.
3—Excessive intake valve clearances.
4—Alternate air door stuck open.

4245. An engine analyzer is an instrument used to

1—measure the fuel/air ratio being burned in the cylinders.
2—monitor the temperature of exhaust gases.
3—measure specific fuel consumption.
4—detect ignition system faults.

4246. A manifold pressure gauge is designed to

1—maintain constant pressure in the intake manifold.
2—indicate differential pressure between the intake manifold and atmospheric pressure.
3—indicate variations of atmospheric pressure at different altitudes.
4—indicate absolute pressure in the intake manifold.

4247. A complete break in the line between the manifold pressure gauge and the induction system of a highly supercharged engine will be indicated by the gauge registering

1—lower than ambient pressure at cruising RPM and higher than ambient pressure at idling RPM.
2—lower than ambient pressure at idling RPM and higher than ambient pressure at cruising RPM.
3—ambient pressure at all engine speeds.
4—higher than ambient pressure at all engine speeds.

4248. The purpose of an exhaust gas analyzer is to indicate the

1—brake specific fuel consumption.
2—fuel/air ratio being burned in the cylinders.
3—temperature of the exhaust gases in the exhaust manifold.
4—grade of fuel being used.

4249. Turbojet engine temperatures are measured by using

1—iron/constantan thermocouples.
2—electrical resistance thermometers.
3—ratiometer electrical resistance thermometers.
4—chromel/alumel thermocouples.

4250. Which of the following types of electric motors are commonly used in electric tachometers?

1—Direct current, series–wound motors.
2—Synchronous motors.
3—Direct current, shunt–wound motors.
4—Direct current, compound–wound motors.

4251. Where are the hot and cold junctions located in an engine cylinder temperature indicating system?

1—Both junctions are located at the instrument.
2—Both junctions are located at the cylinder.
3—The hot junction is located at the cylinder and the cold junction is located at the instrument.
4—The cold junction is located at the cylinder and the hot junction is located at the instrument.

4252. Basically, the indicator of a tachometer system is responsive to change in

1—current flow.
2—voltage polarity.
3—frequency.
4—voltage amplitude.

4253. Which statement is correct concerning a thermocouple–type temperature indicating instrument system?

1—It is a balanced–type, variable resistor circuit.
2—It requires no external power source.
3—It usually contains a balancing circuit in the instrument case to prevent fluctuations of the system voltage from affecting the temperature reading.
4—It will not indicate a true reading if the system voltage varies beyond the range for which it is calibrated.

4254. Which statement is true regarding a thermocouple–type cylinder head temperature measuring system?

1—The resistance required for cylinder head temperature indicators is measured in farads.
2—The voltage output of a thermocouple system is determined by the temperature difference between the two ends of the thermocouple.
3—If a resistor is installed in a thermocouple lead, it is placed in the positive lead.
4—When the master switch is turned on, a thermocouple indicator will move off–scale to the low side.

4255. What basic meter is used to indicate cylinder head temperature in most aircraft?

1—Iron–vane meter.
2—Electrodynamometer.
3—Galvanometer.
4—Thermocouple–type meter.

4256. Which of the following is a primary engine instrument?

1—Tachometer.
2—Torque meter.
3—Fuel flowmeter.
4—Airspeed indicator.

4257. A complete break in the line between the manifold pressure gauge and the induction system will be indicated by the gauge registering

1—prevailing atmospheric pressure.
2—zero.
3—higher than normal for conditions prevailing.
4—lower than normal for conditions prevailing.

4258. Engine oil temperature gauges indicate the temperature of the oil

1—entering the oil cooler.
2—entering the engine.
3—in the oil storage tank.
4—in the return lines to the oil storage tank.

4259. Why do helicopters require a minimum of two synchronous tachometer systems?

1—One indicates engine RPM and the other tail rotor RPM.
2—One indicates main rotor RPM and the other tail rotor RPM.
3—One indicates engine RPM and the other main rotor RPM.
4—Only helicopters with turbine engines employing a dual compressor require two systems.

4260. If the thermocouple leads were inadvertently crossed at installation, what would the cylinder temperature gauge pointer indicate?

1—Normal temperature for prevailing condition.
2—Oscillating pointer.
3—Moves off–scale on the zero side of the meter.
4—Moves off–scale on the high side of the meter.

4261. The instrument used to check heat sensitive elements and heat sensitive bulb resistance is

1—Wheatstone–bridge meter.
2—standing wave ratio (SWR) meter.
3—thermocouple–type meter.
4—meggermeter.

4262. Cylinder head temperatures are measured by the use of a thermocouple circuit which measures the

1—resistance in a metal gasket.
2—difference in the resistance between two dissimilar metals used in the circuit between the hot and cold junctions.
3—difference in the voltage between two metal gaskets.
4—difference in the voltage between two dissimilar metal gaskets.

4263. (1) Engine instruments are color–coded to direct attention to approaching operating difficulties.

(2) Engine instruments are color–coded to pictorially present operating data.

Regarding the above statements, which of the following is true?

1—Both No. 1 and No. 2 are true.
2—Neither No. 1 nor No. 2 is true.
3—Only No. 1 is true.
4—Only No. 2 is true.

4264. Thermocouple leads

1—may be adjusted in length to fit any installation.
2—may be installed with either lead to either post of the indicator.
3—are designed for a specific installation and may not be altered.
4—may be repaired using solderless connectors.

4265. (1) EPR (engine pressure ratio) is a ratio of the engine inlet air pressure to the exhaust gas pressure, and indicates the thrust produced.

(2) EPR (engine pressure ratio) is a ratio of the engine inlet air pressure to the exhaust gas pressure, and indicates volumetric efficiency.

Regarding the above statements, which of the following is true?

1—Only No. 2 is true.
2—Both No. 1 and No. 2 are true.
3—Only No. 1 is true.
4—Neither No. 1 nor No. 2 is true.

4266. What unit in a tachometer system sends information to the indicator?

1—The three–phase a.c. generator.
2—The two–phase a.c. generator.
3—The synchronous motor.
4—The miniature d.c. motor.

4267. Which instrument on a jet engine is used to determine engine power?

1—Turbine inlet temperature gauge.
2—Compressor RPM gauge.
3—Engine pressure ratio gauge.
4—Exhaust gas temperature gauge.

4268. Engine pressure ratio is determined by

1—multiplying engine inlet total pressure by turbine outlet total pressure.
2—multiplying turbine outlet total pressure by engine inlet total pressure.
3—dividing turbine outlet total pressure by engine inlet total pressure.
4—dividing engine inlet total pressure by turbine outlet total pressure.

4269. Jet engine thermocouples are usually constructed of

1—iron–chromel.
2—chromel–alumel.
3—iron–constantan.
4—alumel–constantan.

4270. Which of the following instrument discrepancies would require replacement of the instrument?

A. Red line missing
B. Pointer loose on shaft
C. Glass cracked
D. Mounting screws loose
E. Case paint chipped
F. Leaking at line B nut
G. Will not zero out
H. Fogged

1—G, H, C, B.
2—F, G, C, B.
3—B, F, H, G.
4—B, E, D, F.

4271. A Bourdon–tube instrument may be used to indicate

A. pressure
B. temperature
C. position
D. quantity

1—A and B.
2—C and D.
3—A and C.
4—B and D.

4272. Which of the following instrument discrepancies could be corrected by an aviation mechanic?

A. Red line missing
B. Pointer loose on shaft
C. Glass cracked
D. Mounting screws loose
E. Case paint chipped
F. Leaking at line B nut
G. Will not zero out
H. Fogged

Select corrective actions.

1—A, D, E, F.
2—A, C, E.
3—A, D, F.
4—C, D, E, F.

4273. Which of the following instrument conditions is acceptable and would not require immediate correction?

A. Red line missing
B. Pointer loose on shaft
C. Glass cracked
D. Mounting screws loose
E. Case paint chipped
F. Leaking at line B nut
G. Will not zero out
H. Fogged

1—A.
2—D.
3—None.
4—E.

4274. What indications would be expected on a manifold pressure gauge if the line between the gauge and the induction system was broken?

1—Maximum indication on the gauge.
2—Atmospheric pressure.
3—Pressure of 14.7″ Hg, anytime at sea level.
4—A higher pressure than pressure existing outside of aircraft.

4275. A change in engine manifold pressure has a direct effect on the

1—piston displacement.
2—compression ratio.
3—valve overlap period.
4—mean effective cylinder pressure.

4276. What instrument on a gas turbine engine should be monitored to minimize the possibility of a "hot" start?

1—RPM indicator.
2—Turbine inlet temperature.
3—Horsepower meter.
4—Torquemeter.

4277. The oil temperature indicator on a gas turbine engine indicates the oil temperature

1—at the inlet of the oil pressure pump.
2—as the oil leaves the oil cooler.
3—at the main bearing cavity.
4—as the oil enters the oil reservoir.

4278. On a turbine engine, with a fixed power lever position, the application of engine anti–icing will result in

1—a decrease in EPR.
2—a false EPR reading.
3—an increase in EPR.
4—a modulation of the EPR.

4279. Engine pressure ratio is the total pressure ratio between the

1—front of the compressor and the rear of the compressor.
2—aft end of the compressor and the aft end of the turbine.
3—front of the compressor and the rear of the turbine.
4—front of the engine inlet and the aft end of the compressor.

4280. What would be the possible cause if a gas turbine engine has high exhaust gas temperature, high fuel flow, and low RPM at all engine power settings?

1—Insufficient electrical power to the instrument buss.
2—Fuel control out of adjustment.
3—Loose or corroded thermocouple probes for the EGT indicator.
4—Turbine damage or loss of turbine efficiency.

4281. Gas turbine engine tachometers are usually

1—driven from the main engine shaft.
2—a direct indication of the accessory drive shaft RPM.
3—driven by the quill shaft which indicates RPM of the turbine.
4—calibrated in percent RPM.

4282. What is the primary purpose of the tachometer on an axial–compressor turbine engine?

1—Monitor engine RPM during cruise conditions.
2—The principle instrument for establishing thrust settings.
3—Monitor engine RPM during starting and to indicate overspeed conditions.
4—Monitor power settings to prevent overtemp.

4283. The engine pressure ratio (EPR) indicator is a direct indication of

1—engine thrust being produced.
2—pressures within the turbine section.
3—pressure ratio between the front and aft end of the compressor.
4—ratio of engine RPM to compressor pressure.

4284. The exhaust gas temperature (EGT) indicator on a gas turbine engine provides a relative indication of the

1—exhaust temperature.
2—temperature of the N_1 compressor.
3—temperature of the exhaust gases as they pass the exhaust cone.
4—turbine inlet temperature.

4285. What instrument indicates the thrust of a gas turbine engine?

1—Torquemeter.
2—Exhaust gas temperature indicator.
3—Turbine inlet temperature indicator.
4—Engine pressure ratio indicator.

4286. In a turbine engine, where is the turbine discharge pressure indicator sensor located?

1—At the aft end of the compressor section.
2—At a location in the exhaust cone that is determined to be subjected to the highest pressures.
3—At the eighth stage bleed air port.
4—Immediately aft of the last turbine stage.

4287. In what units are turbine engine tachometers calibrated?

1—Percent of engine RPM.
2—Actual engine RPM.
3—Pounds per square inch (PSI).
4—Percent of engine pressure ratio.

4288. Which of the following fire detectors is commonly used in the power section of an engine nacelle?

1—CO detectors.
2—Combustible mixture detectors.
3—Smoke detectors.
4—Rate-of-temperature-rise detectors.

4289. What is the function of a fire detection system?

1—To discharge the powerplant fire extinguishing system at the origin of the fire.
2—To warn of the presence of fire in the rear section of the powerplant.
3—To activate a warning device in the event of a powerplant fire.
4—To identify the location of a powerplant fire.

4290. Using the chart in Figure 1, determine the fire extinguisher container pressure limits when the temperature is −15° F.

1—125 minimum and 210 maximum.
2—105 minimum and 188 maximum.
3—115 minimum and 199 maximum.
4—115 minimum and 198 maximum.

CONTAINER PRESSURE VERSUS TEMPERATURE

Temperature ° F.	Container Pressure (PSIG)	
	Minimum	Maximum
-40	60	145
-30	83	165
-20	105	188
-10	125	210
0	145	230
10	167	252
20	188	275
30	209	295
40	230	317
50	255	342
60	284	370
70	319	405
80	356	443
90	395	483
100	438	523

Figure 1

4291. How are most aircraft turbine engine fire-extinguishing systems activated?

1—Electrically discharged cartridges.
2—Manual remote control valve.
3—Piston stem and plunger.
4—Pushrod assembly.

4292. How does carbon dioxide extinguish an engine fire?

1—The spray lowers the temperature to a point where combustion will not take place.
2—The spray liquifies in the heat and smothers the fire by shutting off the oxygen supply.
3—The high-pressure spray flushes the fire from the engine.
4—Contact with the air converts the liquid into a gas and snow which smothers the flames.

4293. What retains the nitrogen charge and fire-extinguishing agent in a high rate of discharge (HRD) container?

1—Breakable disk and fusible disk.
2—Pressure switch and check tee valve.
3—Pressure gauge and cartridge.
4—Discharge plug body and strainer.

4294. A continuous-loop fire detector is what type of detector?

1—Spot detector.
2—Overheat detector.
3—Rate-of-temperature-rise detector.
4—Radiation sensing detector.

4295. What is the operating principle of the spot detector sensor in a fire detection system?

1—Resistant core material that prevents current flow at normal temperatures.
2—Fuse material that melts at high temperature.
3—A conventional thermocouple that produces a current flow.
4—A bimetallic thermoswitch that closes when heated to a high temperature.

4296. How is the fire-extinguishing agent distributed in the engine section?

1—Perforated tubing and slinger rings.
2—Spray nozzles and fluid pumps.
3—Nitrogen pressure and slinger rings.
4—Spray nozzles and perforated tubing.

4297. Which of the following is the safest fire-extinguishing agent to use from a standpoint of toxicity and corrosion hazards?

1—Carbon dioxide.
2—Methyl bromide.
3—Bromochloromethane.
4—Water.

4298. Which of the following is not used to detect fires in reciprocating engine nacelles?

1—Smoke detectors.
2—Overheat detectors.
3—Rate-of-temperature-rise detectors.
4—Flame detectors.

4299. What is the principle of operation of the continuous–loop fire detector system sensor?

1—Fuse material which melts at high temperatures.
2—Core resistance material which prevents current flow at normal temperatures.
3—A conventional thermocouple which produces a current flow.
4—A bimetallic thermoswitch which closes when heated to a high temperature.

4300. The most satisfactory extinguishing agent for a carburetor or intake fire is

1—carbon dioxide.
2—dry chemical.
3—methyl bromide.
4—carbon tetrachloride.

4301. The explosive cartridge in the discharge valve of a fire–extinguisher container is

1—a life–dated unit.
2—not a life–dated unit.
3—interchangeable between bottles.
4—mechanically fired.

4302. Why does the Fenwal fire detection system use spot detectors wired in parallel between two separate circuits?

1—A control unit is used to isolate the bad system in case of malfunction.
2—This installation is equal to two systems: a prime system and a reserve system.
3—The dual terminal thermoswitch is used so that one terminal is wired to a bell, the other to a light.
4—A short may exist in either circuit without causing a false fire warning.

4303. Which of the following fire detection systems measures temperature rise compared to a reference temperature?

1—Fenwal continuous loop.
2—Thermocouple.
3—Thermal switch.
4—Lindberg continuous element.

4304. Actuation of an engine fire handle accomplishes what sequence of events?

1—Closes all firewall shutoff valves, disconnects generator, and discharges fire bottle.
2—Closes fuel shutoff, closes hydraulic shut–off, disconnects the generator from the electrical system, and arms the fire–extinguishing system.
3—Silences the fire bell, closes the fuel shutoff valve to the engine, closes the hydraulic shutoff valve, disconnects the generator from the electrical system, and arms the fire–extinguishing system.
4—Closes the fuel shutoff valve to the engine, closes the hydraulic shutoff valve, disconnects the generator from the electrical system, extinguishes the fire warning light, and discharges the fire–extinguishing system into the engine.

4305. A fire detection system operates on the principle of a buildup of gas pressure within a tube proportional to temperature. Which of the following systems does this statement define?

1—Thermocouple fire warning system.
2—Kidde continuous–loop system.
3—Lindberg continuous–element system.
4—Thermal switch system.

4306. The fire detection system that uses gas pressure to close a switch is the

1—Fenwal system.
2—Lindberg system.
3—Kidde system.
4—thermocouple system.

4307. The fire detection system that uses a single wire surrounded by a continuous string of ceramic beads in a tube is the

1—Fenwal system.
2—Lindberg system.
3—Kidde system.
4—thermocouple system.

4308. The fire detection system that uses two wires imbedded in a ceramic core within a tube is the

1—Fenwal system.
2—Lindberg system.
3—thermocouple system.
4—Kidde system.

4309. A gasoline or oil fire is defined as a

1—class A fire.
2—class B fire.
3—class D fire.
4—class C fire.

4310. A fire detection system that operates on the rate of temperature rise is a

1—continuous–loop system.
2—thermocouple system.
3—thermal switch system.
4—continuous element detector.

4311. A fire in an electrical junction box is defined as a

1—class B fire.
2—class A fire.
3—class D fire.
4—class C fire.

4312. Two continuous–loop fire detection systems that will not test due to a broken detector element are the

1—Kidde system and the Lindberg system.
2—Kidde system and the Fenwal system.
3—thermocouple system and the Lindberg system.
4—Kidde system and the thermocouple system.

4313. A high–pressure bottle in an aircraft fire–extinguishing system is stamped "ICC 1–70" or "DOT 1–70" near the neck of the bottle. The stamp indicates that the bottle is ICC or DOT approved and

1—is due to be replaced on 1–70.
2—was manufactured on 1–70.
3—was hydrostatically checked on 1–70.
4—is due for a weight check on 1–70.

4314. In a fixed fire–extinguishing system, there are two small lines running from the system and exiting overboard. These line exit ports are covered with a blowout type indicator disc. Which of the following statements is true?

1—When the yellow indicator disc is missing, it indicates the fire–extinguishing system has had a thermal discharge.
2—When the red indicator disc is missing, it indicates the fire–extinguishing system has been normally discharged.
3—When the yellow indicator disc is missing, it indicates the fire–extinguishing system has been normally discharged.
4—When the green indicator disc is missing, it indicates the fire–extinguishing system has had a thermal discharge.

4315. The most satisfactory extinguishing agent for an electrical fire is

1—water.
2—carbon tetrachloride.
3—carbon dioxide.
4—methyl bromide.

4316. Which of the following fire detection systems will detect a fire when an element is inoperative but will not test when the test circuit is energized?

1—The thermal system and the thermocouple system.
2—The Kidde system and the thermocouple system.
3—The Kidde system and the Fenwal system.
4—The thermocouple system and the Lindberg system.

4317. Which of the following fire detection systems uses heat in the normal testing of the system?

1—The thermocouple system and the Lindberg system.
2—The Kidde system and the Fenwal system.
3—The thermocouple system and the Fenwal system.
4—The Kidde system and the thermocouple system.

4318. What device is used to convert alternating current, which has been induced into the loops of the rotating armature of a d.c. generator, to direct current?

1—An alternator.
2—A rectifier.
3—A commutator.
4—An inverter.

4319. A direct current series motor mounted within an aircraft draws more amperes during start than when it is running under its rated load. The most logical conclusion that may be drawn is

1—the starting winding is shorted.
2—the brushes are floating at operating RPM because of weak brush springs.
3—the condition is normal for this type of motor.
4—hysteresis losses have become excessive through armature bushing (or bearing) wear.

4320. The stationary field strength in a direct–current generator is varied

1—by the reverse–current relay.
2—because of generator speed.
3—because of the number of rotating armature loops available.
4—according to the load requirements.

4321. What type electric motor is generally used with a direct–cranking engine starter?

1—Direct current, shunt–wound motor.
2—Direct current, series–wound motor.
3—Direct current, compound–wound motor.
4—Synchronous motor.

4322. Upon what does the output frequency of an a.c. generator (alternator) depend?

1—The speed of rotation and the strength of the field.
2—The strength of the field and the number of field poles.
3—The speed of rotation, the strength of the field, and the number of field poles.
4—The speed of rotation and the number of field poles.

4323. There is a high surge of current required when a d.c. electric motor is first started. As the speed of the motor increases,

1—the counter e.m.f. decreases proportionally.
2—the applied e.m.f. increases proportionally.
3—the net counter e.m.f. increases until its value is greater than the applied e.m.f.
4—the counter e.m.f. builds up and opposes the applied e.m.f., thus reducing the current flow through the armature.

4324. Alternators (a.c. generators) are often driven by a constant–speed drive mechanism to permit a nearly constant

1—voltage output.
2—amperage output.
3—number of cycles per second.
4—total power output.

4325. What is used to polish commutators or slip rings?

1—Fine emery cloth.
2—Very fine sandpaper.
3—Crocus cloth or fine oilstone.
4—Aluminum oxide or garnet paper.

4326. If the generator is malfunctioning, its voltage can be reduced to residual by actuating the

1—rheostat.
2—master solenoid.
3—overvoltage circuit breaker.
4—master switch.

4327. If the points in a vibrator type voltage regulator stick in the closed position while the generator is operating, what will be the probable result?

1—Generator output voltage will decrease.
2—Generator output voltage will not be affected.
3—Generator output voltage will increase.
4—The reverse–current cutout relay will remove the generator from the line.

4328. Why is a constant–speed drive used to control the speed of some aircraft engine–driven generators?

1—So that the voltage output of the generator will remain within limits.
2—To eliminate uncontrolled surges of current to the electrical system.
3—So that both voltage and amperage output can be controlled directly.
4—So that the frequency of the alternating current output will remain constant.

4329. What will be the result if a short circuit occurs between the positive field lead and the positive armature lead of a generator with the engine operating at cruising RPM?

1—The reverse–current cutout relay will not close.
2—A high generator voltage.
3—Failure of the generator to produce any voltage.
4—The generator will only produce residual voltage.

4330. Aircraft that operate more than one generator connected to a common electrical system must be provided with

1—automatic generator switches that operate to isolate any generator whose output is less than 80 percent of its share of the load.
2—an automatic device that will isolate nonessential loads from the system if one of the generators fails.
3—a generator switch arrangement that will prevent any one generator from being connected to the system unless the other generators are operating.
4—individual generator switches that can be operated from the cockpit during flight.

4331. The most effective method of regulating aircraft direct current generator output is to vary, according to the load requirements, the

1—strength of the stationary field.
2—generator speed.
3—effective resistance in the load circuit.
4—number of rotating armature loops in use.

4332. Electric motors are often classified according to the method of connecting the field coils and armature. Aircraft engine starter motors are generally of which type?

1—Compound.
2—Series.
3—Differential compound.
4—Shunt (parallel).

4333. As the generator load is increased (within its rated capacity), the voltage will

1—decrease and the amperage output will increase.
2—increase and the amperage output will increase.
3—remain constant and the amperage output will increase.
4—remain constant and the amperage output will decrease.

4334. As the flux density in the field of a d.c. generator increases and the current flow to the system increases,

1—the force required to turn the generator decreases.
2—the generator voltage decreases.
3—the generator amperage decreases.
4—the force required to turn the generator increases.

4335. Alternators used on light aircraft require which of the following control devices?

1—Speed control regulator.
2—Reverse–current cutout relay.
3—Voltage regulator.
4—Current regulator.

4336. What is the purpose of a reverse–current cutout relay?

1—It eliminates the possibility of reversed polarity of the generator output current.
2—It prevents overloading the generator.
3—It prevents fluctuations of generator voltage.
4—It opens the main generator circuit whenever the generator voltage drops below the battery voltage.

4337. Generator voltage will not build up when the field is flashed and solder is found on the brush cover plate. These are most likely indications of

1—an open armature.
2—a sheared armature shaft.
3—a shorted armature.
4—damaged armature shaft bearings.

4338. Why is it unnecessary to flash the field of the exciter on a brushless alternator?

1—The exciter is constantly charged by battery voltage.
2—Brushless alternators do not have exciters.
3—Permanent magnets are installed in the main field poles.
4—The slip ring employed by the brushless alternator retains a permanent charge.

4339. How is the automatic ignition relight switch activated on a gas turbine engine?

1—By a sensing switch located in the tailpipe.
2—By a decrease in tailpipe temperature.
3—By a drop in the compressor–discharge pressure.
4—By a drop in fuel flow.

4340. How are the rotor windings of an aircraft alternator usually excited?

1—By a constant a.c. voltage from the battery.
2—With alternating current from a permanent condensor.
3—By a constant a.c. voltage.
4—By a variable direct current.

4341. What precaution is usually taken to prevent electrolyte from freezing in a lead acid battery?

1—Place the aircraft in a hangar.
2—Remove the battery and place it in a warm area.
3—Keep the battery fully charged.
4—Drain the electrolyte.

4342. What is the ampere–hour rating of a storage battery that is designed to deliver 45 amperes for 2.5 hours?

1—112.5 ampere–hour.
2—47.5 ampere–hour.
3—90.0 ampere–hour.
4—45.0 ampere–hour.

4343. How many hours will a 140 ampere–hour battery deliver 15 amperes?

1—15.0 hours.
2—1.40 hours.
3—9.33 hours.
4—14.0 hours.

4344. Which of the following aircraft circuits does not contain a fuse?

1—Generator circuit.
2—Air–conditioning circuit.
3—Exterior lighting circuit.
4—Starter circuit.

4345. The maximum number of terminals that may be connected to any one terminal stud in an aircraft electrical system is

1—four.
2—one.
3—two.
4—three.

4346. What is the maximum number of bonding jumper wires that may be attached to one terminal grounded to a flat surface?

1—Four.
2—Five.
3—Two.
4—Three.

4347. As a general rule, starter brushes are replaced when they are

1—approximately one–half their original length.
2—approximately one–fourth their original length.
3—1/8–inch long.
4—1/2–inch long.

4348. When installing an electrical switch, under which of the following conditions should the switch be derated from its nominal current rating?

1—Conductive circuits.
2—Capacitive circuits.
3—Low rush–in circuits.
4—Direct–current motor circuits.

4349. The resistance of the current return path through the aircraft is always considered negligible, provided the

1—voltage drop across the circuit is checked.
2—circuit resistance is checked.
3—generator is properly grounded.
4—structure is adequately bonded.

4350. When should parallel electrical wires be twisted on installation?

1—When desired to keep the bundle rigid.
2—When not tied in bundles.
3—When necessary to reduce the wire diameter.
4—When used in the vicinity of a magnetic compass.

4351. When does current flow through the coil of a solenoid–operated electrical switch?

1—Continually, as long as the aircraft's electrical system master switch is on.
2—Continually, as long as the control circuit is complete.
3—Only for a short time period following movement of the control switch.
4—Only until the movable points contact the stationary points.

4352. It is necessary to determine that the electrical load limit of a 28–volt, 75–amp generator, installed in a particular aircraft, has not been exceeded. By making a ground check, it is determined that the battery furnished 57 amperes to the system when all equipment that can continuously draw electrical power in flight is turned on. This type of load determination

1—cannot be made because generator capacity exceeds 2–1/2 kW.
2—can be made, but the load will exceed the generator load limit.
3—can be made, and the load will be within the generator load limit.
4—cannot be made on direct current electrical systems.

4353. What type of lubricant may be used to aid in pulling electrical wires or cables through conduits?

1—Lightweight, vegetable–base grease.
2—Powdered graphite.
3—Soapstone talc.
4—Rubber lubricant.

4354. Bonding jumpers should be designed and installed in such a manner that they

1—are not subjected to flexing by relative motion of airframe or engine components.
2—limit the relative motion of the parts to which they are attached by acting as a secondary stop.
3—provide a low electrical resistance in the ground circuit.
4—prevent buildup of a static electrical charge between the airframe and the surrounding atmosphere.

4355. On a turbine engine, with the starter–generator circuit energized, the engine would not crank. The probable cause would be the

1—overvoltage relay is defective.
2—throttle ignition switch is defective.
3—igniter relay is defective.
4—starter relay is defective.

4356. Arcing at the brushes and burning of the commutator of a motor may be caused by

1—weak brush springs.
2—excessive brush spring tension.
3—smooth commutator.
4—low–mica.

4357. The maximum allowable voltage drop between the generator and the bus bar is

1—1 percent of the regulated voltage.
2—greater than the voltage drop permitted between the battery and the bus bar.
3—2 percent of the regulated voltage.
4—less than the voltage drop permitted between the battery and the bus bar.

4358. Switches used to control engine electrical circuits should be installed

1—upside down to prevent debris from shorting the terminals.
2—so the toggle will move in the same direction as the desired motion of the unit controlled.
3—under a guard.
4—so the ON position is reached by a forward or upward position.

4359. When selecting an electrical switch for installation in an aircraft circuit utilizing a direct current motor,

1—a switch designed for d.c. should be chosen.
2—the switch must be a single pole single throw (SPST).
3—a derating factor should be applied.
4—only switches with screw type terminal connections should be used.

4360. When installing electrical wiring parallel to a fuel line, the wiring should be

1—in a metal conduit.
2—in a vinyl sleeve.
3—above the fuel line.
4—below the fuel line.

4361. If an electrical unit which requires 28 volts and 40–ampere flow is to be installed using No. 4 copper cable, how many feet from the main distribution source can the unit be located without exceeding a 1–volt line drop? (See Figure 2.)

1—13 feet.
2—90 feet.
3—6.5 feet.
4—100 feet.

4362. In a 28–volt system, what is the maximum continuous current that can be carried by a single No. 10 copper wire 25 feet long, routed in free air? (See Figure 2.)

1—15 amperes.
2—20 amperes.
3—35 amperes.
4—28 amperes.

4363. What is the basic unit of measure for capacitance?

1—Henry.
2—Gauss.
3—Ohm.
4—Farad.

ELECTRIC WIRE CHART

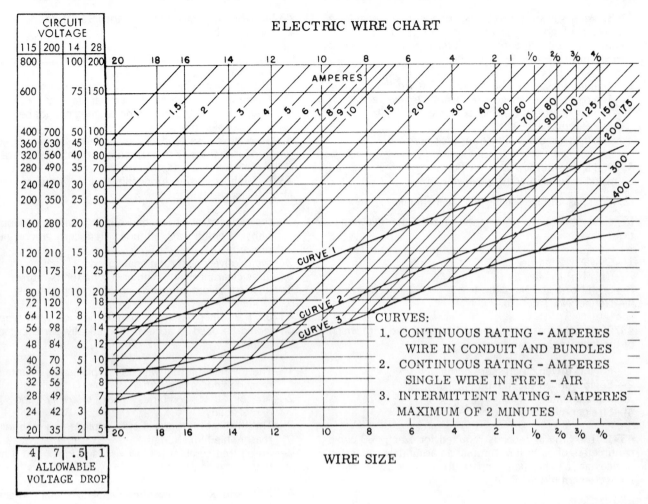

Figure 2

4364. If the voltage applied across the plates of a capacitor (condenser) is too great

1—the plates will become saturated and not accept an electrical charge.
2—the dielectric will break down and arcing will occur between the plates.
3—the induced counter EMF will cause the capacitor to act as a resistor in the circuit.
4—it will probably not damage the capacitor because there is no physical connection between the plates; however, it may cause some secondary problems in other parts of the circuit.

4365. In an alternating current circuit, the effective voltage

1—is equal to the maximum instantaneous voltage.
2—is less than the maximum instantaneous voltage.
3—is greater than the maximum instantaneous voltage.
4—may be greater than or less than the maximum instantaneous voltage.

4366. The amount of electricity a capacitor can store is directly proportional to

1—the plate area and inversely proportional to the distance between the plates.
2—the distance between the plates and inversely proportional to the plate area.
3—the plate area and is not affected by the distance between the plates.
4—the distance between the plates and is not affected by the plate area.

4367. When capacitors are connected in parallel, the total capacitance is equal to

1—the sum of the reciprocals of the capacitances.
2—the sum of the capacitances divided by the number of capacitors.
3—the reciprocal of the sum of the reciprocals of the capacitances.
4—the sum of the capacitances.

4368. What will be the result of operating an engine in extremely high temperatures using a lubricant recommended by the manufacturer for a much lower temperature?

1—The oil pressure will be higher than normal.
2—The oil pressure gauge will fluctuate excessively.
3—The oil temperature and oil pressure will be higher than normal.
4—The oil pressure will be lower than normal.

4369. (1) Gas–turbine and reciprocating engine oils can be mixed or used interchangeably.

(2) Most gas–turbine engine oils are synthetic.

Regarding the above statements, which of the following is true?

1—Only No. 2 is true.
2—Only No. 1 is true.
3—Both No. 1 and No. 2 are true.
4—Neither No. 1 nor No. 2 is true.

4370. An oil separator is generally associated with which of the following?

1—Engine–driven oil pressure pump.
2—Engine–driven vacuum pump.
3—Cuno oil filter.
4—Strainer–type filter.

4371. The time in seconds required for exactly 60 cubic centimeters of oil to flow through an accurately calibrated orifice at a specific temperature is recorded as a measurement of the oil's

1—flash point.
2—specific gravity.
3—viscosity.
4—pour point.

4372. Upon what quality or characteristic of a lubricating oil is its viscosity index based?

1—Its ability to maintain film strength.
2—Its resistance to flow at a standard temperature as compared to high grade paraffin–base oil at the same temperature.
3—Its rate of change in viscosity with temperature change.
4—Its rate of flow through an orifice at a standard temperature.

4373. Lubricating oils with high viscosity index ratings are oils

1—in which the viscosity does not vary much with temperature change.
2—in which the viscosity varies considerably with temperature change.
3—which have high pour points.
4—which have high SAE numbers.

4374. Why are synthetic lubricants used in high–performance turbine engines?

1—Synthetic oils do not require filtering and are less expensive.
2—The load–carrying characteristics of petroleum–base oils have a low degree of chemical stability.
3—Additives required in turbine engines cannot be mixed with petroleum oils.
4—They have less tendency to produce lacquer or coke and less tendency to evaporate at high temperatures.

4375. High viscosity lubricating oil is used in most aircraft engines due to

1—the reduced ability of thin oils to maintain adequate film strength at altitude (reduced atmospheric pressure).
2—the relatively high rotational speeds.
3—large clearances and high operating temperatures.
4—its lower oxidation rate at elevated temperatures.

4376. If all other requirements can be met, what type of oil should be used to achieve theoretically perfect engine lubrication?

1—The thinnest oil that will stay in place and maintain a reasonable film strength.
2—An oil that combines high viscosity and low demulsibility.
3—The thickest oil that will stay in place and maintain a reasonable film strength.
4—An oil that combines a low viscosity index and a high neutralization number.

4377. In addition to preventing metal–to–metal contact of moving parts, an engine lubricant aids in how many of the following?

(1) Cooling
(2) Sealing
(3) Cleaning
(4) Corrosion prevention

1—One.
2—Four.
3—Two.
4—Three.

4378. The type of lubricating oil that is used in a turbine aircraft engine is

1—synthetic.
2—petrolatum.
3—50–50 blend of petroleum and synthetic.
4—30–70 blend of petroleum and synthetic.

4379. When selecting an engine oil, which of the following characteristics is important?

1—High pour point.
2—Low flash point.
3—High flash point.
4—Low degree of chemical stability.

4380. The viscosity of a liquid is a measure of its

1—resistance to flow.
2—rate of change of internal friction with change in temperature.
3—density.
4—ability to transmit force.

4381. What type of oil system is usually found on turbojet engines?

1—Dry sump, pressure, and spray.
2—Wet sump, dip, and pressure.
3—Dry sump, dip, and splash.
4—Wet sump, spray, and splash.

4382. The engine's lubricating oil aids in reducing friction, cushioning shock, and in

1—cooling the engine.
2—preventing fatigue of engine parts.
3—heating fuel in carburetor to prevent ice.
4—preventing a buildup of internal pressures in the crankcase.

4383. Which of the following factors determines the proper grade of oil to use in a particular engine?

1—High viscosity to provide good flow characteristics.
2—Adequate lubrication in various attitudes of flight.
3—Positive introduction of oil to the bearings.
4—Operating speeds of bearings.

4384. Specific gravity is a comparison of the weight of a substance to the weight of an equal volume of

1—oil at a specific temperature.
2—distilled water at a specific temperature.
3—mercury at a specific temperature.
4—isopropyl at a specific temperature.

4385. Which of the following has the greatest effect on the viscosity of lubricating oil?

1—Temperature.
2—Lubricity.
3—Pressure.
4—Volatility.

4386. What advantage do mineral base lubricants have over vegetable oil base lubricants when used in aircraft engines?

1—Cooling ability.
2—Sealing quality.
3—Chemical stability.
4—Friction resistance.

4387. Lubricants may be classified according to their origin. Satisfactory aircraft engine lubricants are

1—mineral or synthetic based.
2—animal, vegetable, mineral, or synthetic based.
3—vegetable, mineral, or synthetic based.
4—animal, mineral, or synthetic based.

4388. What are the functions of the lubricating oil in an aircraft engine?

1—Lubricates, cools, cleans, and prevents fatigue of parts.
2—Lubricates, cools, seals, and prevents internal pressure buildup.
3—Lubricates, seals, cools, and cleans.
4—Lubricates and increases friction between moving parts.

4389. What type of oil pump is most commonly used on turbojet engines?

1—Gear.
2—Centrifugal.
3—Vane.
4—Diaphragm.

4390. If a high-powered engine has been ground operated at high power output long enough to attain high operating temperatures, the mechanic should be careful not to decelerate the engine too quickly due to the possiblilty of

1—carbonizing the oil trapped in the ring grooves.
2—rupturing the diaphragm control valve in the automatic oil temperature control unit.
3—overloading the power section with oil.
4—completely scavenging all power section oil.

4391. The engine oil temperature regulator is usually located between which of the following on a dry sump engine?

1—The engine oil supply pump and the internal lubrication system.
2—The scavenger pump outlet and the oil storage tank.
3—The oil storage tank and the engine oil supply pump.
4—The sumps and the scavenger pump inlet.

4392. What will happen to the return oil if the oil line between the scavenger pump and the oil cooler separates?

1—Oil will accumulate in the engine.
2—The return oil will be pumped overboard.
3—The cooler check valve will close and force the oil to bypass the cooler core and return to the tank via the cold oil line.
4—The scavenger return line check valve will close and force the oil to bypass directly to the intake side of the pressure pump.

4393. At cruise RPM, some oil will flow through the relief valve of a gear-type engine oil pump. This is normal as the relief valve is set at a pressure which is

1—lower than the pump inlet pressure.
2—lower than the pressure pump capabilities.
3—higher than pressure pump capabilities.
4—lower than the scavenger pump capabilities.

4394. (1) Fuel may be used to cool oil in gas–turbine engines.

(2) Ram air may be used to cool oil in gas–turbine engines.

Regarding the above statements, which of the following is true?

1—Only No. 1 is true.
2—Only No. 2 is true.
3—Neither No. 1 nor No. 2 is true.
4—Both No. 1 and No. 2 are true.

4395. In the oil system, where is the temperature bulb normally located?

1—At the oil supply tank inlet.
2—In the scavenger pump inlet.
3—At the Y drain in all engines.
4—In the pressure oil screen housing.

4396. In a newly installed engine, if the oil pump runs but will not pump oil, the supply of oil is sufficient, and there are no leaks in the oil lines, then

1—the pump has excessive side clearance.
2—an air lock in the relief valve is indicated.
3—the pressure relief valve is stuck closed.
4—the pump should be replaced with a new one.

4397. If a thermostatic–type oil temperature control valve should fail in the open position, the oil would

1—bypass the radiator core and return to the inlet of the scavenger pump.
2—flow through the radiator core and return to the supply tank.
3—bypass the radiator core and return to the supply tank.
4—flow through the radiator core and return to the inlet of the scavenger pump.

4398. What is the purpose of the last chance oil filters?

1—To allow the oil to bypass the main filter in the event it becomes clogged.
2—To prevent damage to the oil spray nozzle.
3—To filter the oil immediately before it enters the main bearings.
4—To assure a clean supply of oil to the lubrication system.

4399. In a jet engine which uses a fuel–oil heat exchanger, the oil temperature is controlled by a thermostatic valve that regulates the flow of

1—air past the heat exchanger.
2—fuel through the heat exchanger.
3—both fuel and oil through the heat exchanger.
4—oil through the heat exchanger.

4400. What prevents pressure within the lubricating oil tank from rising above or falling below ambient pressure (reciprocating engine)?

1—The oil tank check valve.
2—The oil pressure relief valve.
3—The oil tank vent.
4—The thermostatic bypass valve.

4401. In an axial–flow turbine engine, compressor bleed–air is sometimes used to aid in cooling the

1—oil.
2—inlet guide vanes.
3—oil cooler.
4—turbine.

4402. Oil picks up the most heat from which of the following turbojet engine components?

1—Rotor coupling.
2—Compressor bearing.
3—Accessory drive bearing.
4—Turbine bearing.

4403. Which of the following is a function of the fuel–oil heat exchanger on a turbojet engine?

1—Removes oil vapors.
2—Aerates the fuel.
3—Emulsifies the oil.
4—Increases fuel temperature.

4404. Oil tank filler openings on turbine engines must be marked with the word

1—"oil" and the type and grade of oil specified by the manufacturer.
2—"oil" and the tank capacity.
3—"capacity" and grade.
4—"type" of oil approved for the engine.

4405. After making a welded repair to an unpressurized aluminum oil tank used in a turbojet aircraft, the tank should be pressure checked to

1—5.0 PSI.
2—3.5 PSI.
3—2.0 PSI.
4—6.5 PSI.

4406. Why are fixed orifice nozzles used in the lubrication system of gas turbine engines?

1—To provide a relatively constant oil flow to the main bearings at all engine speeds.
2—To keep back pressure on the oil pump, thus preventing an air lock.
3—To protect the oil seals by preventing excessive pressure from entering the bearing cavities.
4—To reduce the oil pressure.

4407. Why are oil coolers not used on turbine engines with wet–sump lubrication systems?

1—External oil tubes placed in selected areas about the engine are cooled by inlet air.
2—Cooling air is directed to the turbine wheel and bearings.
3—The oil tank serves as both a reservoir and cooler.
4—Synthetic oil runs cooler than mineral base oil eliminating the need for additional cooling.

4408. What would be the probable result if the oil system pressure relief valve should stick in the open position on a turbine engine?

1—Increased oil pressure.
2—Decreased oil temperature.
3—Insufficient lubrication.
4—Pressurization of the case and increased oil leakage.

4409. What is the primary purpose of the oil–to–fuel heat exchanger?

1—Cool the fuel.
2—Cool the oil.
3—De–aerate the oil.
4—Decrease the viscosity of the oil.

4410. What unit in an aircraft engine lubrication system is adjusted to maintain the desired system pressure?

1—Oil pressure relief valve.
2—Oil filter bypass valve.
3—Oil pump.
4—Oil pressure indicator.

4411. Low oil pressure can be detrimental to the internal engine components. However, high oil pressure

1—is desirable for maximum bearing life.
2—should be limited to the engine manufacturer's recommendations.
3—is not important and may be ignored.
4—will not occur because of pressure losses around the bearings.

4412. What is the primary purpose of the oil breather pressurization system that is used on turbine engines?

1—Positive pressure prevents contaminants from entering the system.
2—Prevents foaming of the oil.
3—Allows aeration of the oil for better lubrication because of the air/oil mist.
4—Provides a proper oil spray pattern from the main bearing oil jets.

4413. What is the purpose of directing bleed air to the bearings on turbine engines?

1—It increases the oil pressure for better lubrication.
2—It provides a high volume of oil to the most critical bearings.
3—It heats the oil to the proper operating temperature.
4—It eliminates the need for an oil cooler in a wet–sump lubrication system.

4414. What is the purpose of an oil cooler bypass valve in a dry sump engine?

1—To direct cold oil into the oil supply tank.
2—To direct cold oil to the intake side of the pressure pump.
3—To direct hot oil into the oil supply tank.
4—To direct cold oil into the oil filter.

4415. In order to relieve excessive pump pressure in an engine's internal oil system, most engines are equipped with a

1—vent.
2—bypass valve.
3—breather.
4—relief valve.

4416. What is the major single source of heat which is carried away by the lubricating oil in a reciprocating aircraft engine?

1—The connecting rod bearings.
2—The cylinder walls.
3—The main rod bearings.
4—The accessory section.

4417. How are the teeth of the gears in the accessory section of an engine normally lubricated?

1—By splashed or sprayed oil.
2—By submerging the load–bearing portions in oil.
3—By surrounding the load–bearing portions with baffles or housings within which oil pressure can be maintained.
4—By pressure oil directed from the gear hub out through the webs (spokes) to the individual teeth.

4418. What is the purpose of the check valve generally used in a dry–sump lubrication system?

1—To prevent the oil from the oil temperature regulator from returning to the crankcase during inoperative periods.
2—To prevent the scavenger pump from losing its prime.
3—To prevent the oil from the supply tank from seeping into the crankcase during inoperative periods.
4—To prevent the oil from the pressure pump from entering the scavenger system.

4419. From the following, identify the factor that has the least effect on the oil consumption of a specific engine.

1—Mechanical efficiency.
2—Engine temperature.
3—Engine RPM.
4—Lubricant characteristics.

4420. How is the oil collected by the piston oil ring returned to the crankcase?

1—Down vertical slots cut in the piston wall between the piston oil ring groove and the piston skirt.
2—Through hollow piston pins.
3—Through holes drilled in the piston oil ring groove.
4—Through holes drilled in the piston pin recess.

4421. Which of the following lubrication system components is never located between the pressure pump and the engine pressure system?

1—Oil temperature bulb.
2—Oil filter or strainer.
3—Fuel line for oil dilution system.
4—Check valve.

4422. As an aid to cold–weather starting, the oil dilution system thins the oil with

1—propane.
2—kerosene.
3—alcohol.
4—gasoline.

4423. The basic oil pressure relief valve setting for a newly overhauled engine is made

1—within the first 30 seconds of engine operation.
2—when the oil is at a higher than normal temperature to assure high oil pressure at normal oil temperature.
3—within 1 minute after the first start.
4—in the overhaul shop.

4424. Where is the oil temperature bulb located on a dry–sump reciprocating engine?

1—Oil inlet line.
2—Oil cooler.
3—Oil outlet line.
4—Oil scavenge sump.

4425. Cylinder walls are usually lubricated by

1—splashed or sprayed oil.
2—a direct pressure system fed through the crankshaft, connecting rods, and the piston pins to the oil control ring groove in the piston.
3—oil that is picked up by the oil control ring when the piston is at bottom center.
4—oil migration past the rings during the intake stroke.

4426. If a full–flow oil filter is used on an aircraft engine, and the filter becomes clogged, the

1—pressure buildup in the filter will collapse the screen and close off the oil supply to the engine.
2—oil will be bypassed to the magnetic oil sump plug where metallic particles will be removed.
3—oil will be bypassed back to the oil tank hopper where sediment and foreign matter will settle out prior to passage through the engine.
4—bypass valve will open and the oil pump will supply unfiltered oil to the engine.

4427. Oil accumulation in the cylinders of an inverted in–line engine and in the lower cylinders of a radial engine is normally reduced or prevented by

1—reversed oil control rings.
2—closing the oil shutoff valve after shutdown.
3—routing the valve–operating mechanism lubricating oil to a separate scavenger pump.
4—extended cylinder skirts.

4428. How can engine roller bearing and gear lubrication best be accomplished?

1—By dip or spray oiling.
2—By submerging the load bearing portion in oil.
3—By operation in a closed housing to which a supply of oil under pressure can be connected.
4—By application of fibrous high–temperature grease.

4429. What is the primary purpose of changing aircraft engine lubricating oils at predetermined periods?

1—Exposure to heat and oxygen causes the oil to lose its ability to maintain a film under load.
2—The oil becomes contaminated with finely divided solid particles suspended in the oil.
3—The oil eventually wears out.
4—The oil eventually becomes diluted due to unvaporized gasoline washing past the pistons and into the crankcase.

4430. What determines the minimum particle size which will be excluded or filtered by a cuno–type (stacked disc, edge filtration) filter?

1—The number of discs in the assembly.
2—The disc thickness.
3—The spacer thickness.
4—Both the number and thickness of the discs in the assembly.

4431. What is the primary purpose of the hopper located in the oil supply tank of some dry–sump engine installations?

1—To reduce the time required to warm the engine to operating temperatures.
2—To reduce surface aeration of the hot oil and thus reduce oxidation and the formation of sludge and varnish.
3—To cause warm oil to mix with the cold oil without stratification and subsequent variation in viscosity.
4—To impart a centrifugal motion to the oil entering the tank so that the foreign particles in the oil will separate more readily.

4432. What is the purpose of the flow control valve in a reciprocating engine oil system?

1—Direct oil through or around the oil cooler.
2—Deliver cold oil to the hopper tank.
3—Relieve excessive pressures in the oil cooler.
4—Compensate for volumetric increases due to foaming of the oil.

4433. Where are sludge chambers, when used in aircraft engine lubrication systems, usually located?

1—In the crankshaft throws.
2—Adjacent to the scavenger pumps.
3—In the oil storage tank.
4—In the crankshaft tail shaft if transfer rings are used.

4434. Why are all oil tanks equipped with vent lines?

1—To prevent pressure buildup in the engine.
2—To eliminate foaming in the tank.
3—To prevent pressure buildup in the tank.
4—To eliminate foaming in the engine.

4435. Excessive oil is prevented from accumulating on the cylinder walls of a reciprocating engine by

1—the design shape of the piston skirt.
2—holes drilled in the piston skirt.
3—internal engine pressure bleeding past the ring grooves.
4—oil control rings on the pistons.

4436. (1) Wet–sump oil systems are most commonly used in gas–turbine engines.

(2) Oil in gas–turbine engines is not diluted during cold weather.

Regarding the above statements, which of the following is true?

1—Only No. 1 is true.
2—Both No. 1 and No. 2 are true.
3—Only No. 2 is true.
4—Neither No. 1 nor No. 2 is true.

4437. The pumping capacity of the scavenger pump in a dry–sump aircraft engine's lubrication system

1—is greater than the capacity of the oil supply pump.
2—is less than the capacity of the oil supply pump.
3—is usually equal to the capacity of the oil supply pump in order to maintain constant oiling conditions.
4—varies according to the oil supply tank capacity and not according to the oil supply pump capacity.

4438. In which of the following situations will the oil cooler automatic bypass valve be open the greatest amount?

1—Engine oil at normal operating temperature.
2—Engine oil above normal operating temperature.
3—Engine oil below normal operating temperature.
4—Engine stopped with no oil flowing after runup.

4439. Which of the following statements in reference to oil tanks is correct?

1—The diameter of the oil tank outlet shall be at least two times the diameter of the engine inlet and in no case less than 1 inch.
2—The word oil and the oil tank capacity shall be marked on or adjacent to the filler cover of all oil tanks.
3—All oil tanks shall be capable of withstanding, without failing, an internal test pressure of 15 PSI.
4—All oil tank outlets shall be provided with a finger screen to prevent the flow of foreign solid matter from the storage tank into the lines.

4440. In order to maintain a constant oil pressure as the clearances between the moving parts of an engine increase through normal wear, the supply pump output must

1—increase as the resistance offered to the flow of oil decreases.
2—decrease as the resistance offered to the flow of oil increases.
3—increase as the resistance offered to the flow of oil increases.
4—decrease as the resistance offered to the flow of oil decreases.

4441. The overhead valve assemblies of opposed engines used in helicopters are lubricated by means of a

1—forced feed system.
2—splash and spray system.
3—pressure system.
4—combination splash and spray and gravity feed system.

4442. What will result if an oil screen becomes completely blocked?

1—Oil will flow at 75 percent of the normal rate through the system.
2—Oil flow to the engine will stop.
3—Oil flow from the engine will stop.
4—Oil will flow at the normal rate through the system.

4443. A turbine engine dry–sump lubrication system of the self–contained, high–pressure design

1—uses the same storage area as a wet–sump engine.
2—has no heat exchanger.
3—consists of pressure, breather, and scavenge subsystems.
4—stores oil in the engine crankcase.

4444. Lube system last chance filters in turbine engines are usually cleaned

1—during annual inspection.
2—during 100–hour inspections.
3—during overhaul.
4—at oil change intervals.

4445. How are the piston pins of most aircraft engines lubricated?

1—By pressure oil through a drilled passageway in the heavy web portion of the connecting rod.
2—By oil which is sprayed or thrown by the master or connecting rods.
3—By the action of the oil control ring and the series of holes drilled in the ring groove directing oil to the pin and piston pin boss.
4—By pressure oil through a drilled passage the entire length of the linkrod.

4446. The vent line connecting the oil supply tank and the engine in some dry–sump engine installations permits

1—pressurization of the oil supply to prevent cavitation of the oil supply pump.
2—oil vapors from the engine to be condensed and drained into the oil supply tank.
3—the oil tank to be vented through the normal engine vent.
4—the engine and oil supply tank to be vented to each other, thus avoiding the use of an atmospheric vent.

4447. An engine lubrication system pressure relief valve is usually located between the

1—oil cooler and the scavenger pump.
2—scavenger pump and the external oil system.
3—pump and the internal oil system.
4—sump and the scavenger pump.

4448. Why is the scavenger pump used in a particular dry–sump system designed with a greater capacity than the pressure pump used in the same system?

1—The scavenger pump oil intake is by suction, whereas the pressure pump oil intake is fed by gravity.
2—The scavenger pump lines are longer than the pressure pump lines.
3—The scavenger pump must handle a greater volume of oil than the pressure pump.
4—The scavenger pump must develop a higher pressure than the pressure pump.

4449. Where is the oil of a dry–sump reciprocating engine exposed to the temperature control valve sensing unit?

1—Oil cooler inlet.
2—Engine outlet.
3—Oil strainer.
4—Engine inlet.

4450. Under which of the following conditions is the oil cooler flow control valve open on a reciprocating engine?

1—When the temperature of the oil returning from the engine is too high.
2—When the engine pump output volume exceeds the scavenger pump output volume.
3—When the temperature of the oil returning from the engine is too low.
4—When the scavenger pump output volume exceeds the engine pump input volume.

4451. Most jet engine oil tanks incorporate a check relief valve in the tank venting system. The purpose of this valve is to

1—prevent oil pump cavitation by maintaining a constant pressure on the oil pump inlet.
2—prevent the return oil from foaming in the tank.
3—prevent loss of oil overboard during aircraft acceleration.
4—prevent oil from draining from the oil tank and flooding the engine sump when the engine is shut down.

4452. In a reciprocating engine, oil is directed from the pressure relief valve to the inlet side of the

1—scavenger pump.
2—thermostatic control.
3—oil temperature regulator.
4—pressure pump.

4453. If oil became congealed in a cooler, what unit would prevent damage to the cooler?

1—Core.
2—Thermostatic control valve.
3—Annular jacket.
4—Baffle plates.

4454. The primary source of oil contamination in a normally operating radial aircraft engine is

1—metallic deposits as a result of engine wear.
2—atmospheric dust and pollen.
3—combustion deposits due to combustion chamber blow–by and oil migration on the cylinder walls.
4—oil decomposition as a result of exposure to oxygen in the air.

4455. A rise in oil temperature and a drop in oil pressure may be caused by

1—the temperature regulator sticking shut.
2—the pressure relief valve sticking shut.
3—foreign material under the relief valve.
4—improper starting procedure, engine not warmed up.

4456. The main oil filters strain the oil at which point in the system?

1—Immediately after it leaves the scavenger pump.
2—Immediately before it enters the pressure pump.
3—Just before it passes through the spray nozzles.
4—Just as it leaves the pressure pump.

4457. Which type valve prevents oil from entering the main accessory case when the engine is not running?

1—Bypass.
2—Relief.
3—Check.
4—Restriction.

4458. A component often found in jet engine dry–sump lubrication systems, but not in wet–sump systems, is the

1—oil cooler.
2—reservoir.
3—pressure pump.
4—scavenge pump.

4459. An oil tank having a capacity of 5 gallons must have an expansion space of

1—3 quarts.
2—2 quarts.
3—4 quarts.
4—5 quarts.

4460. Oil tank flapper valves are most likely to be closed when the aircraft is

1—flying at a constant speed.
2—taking off.
3—increasing speed.
4—decreasing speed.

4461. Why is expansion space required in an engine oil supply tank?

1—To eliminate oil foaming.
2—For oil enlargement and collection of foam.
3—To ensure gravity oil feed.
4—For proper oil tank ventilation.

4462. The air and oil are separated in a jet engine oil system by returning the scavenged oil to

1—a centrifugal separator.
2—the bottom of the reservoir.
3—a pressurized tank.
4—a de-aerator at the top of the reservoir.

4463. Which of the following bearing types must be continuously lubricated by pressure oil?

1—Ball.
2—Roller.
3—Tapered.
4—Friction.

4464. When a magneto is disassembled, keepers are usually placed across the poles of the rotating magnet to reduce the loss of magnetism. These keepers are usually made of

1—chrome magnet steel.
2—soft iron.
3—cobalt steel.
4—laminated high-carbon steel.

4465. How is the strength of a magneto magnet checked?

1—Hold the points open and check the output of the primary coil with an a.c. ammeter while operating the magneto at a specified speed.
2—Check the a.c. voltage reading at the breaker points.
3—Check the output of the secondary coil with an a.c. ammeter while operating the magneto at a specified speed.
4—While operating the magneto at any speed, determine the size of air gap the spark from the magneto will jump.

4466. What is the dwell angle of a magneto?

1—The angle between full register and neutral position.
2—The distance in degrees the cam travels while the breaker points are open.
3—The angle at which the rotating magnet and the field windings produce their highest voltage.
4—The angle between breaker cam lobes during which the points are closed.

4467. The E-gap angle is usually defined as the number of degrees between the neutral position of the rotating magnet and the position

1—where the contact points close.
2—where the contact points open.
3—of greatest magnetic flux density.
4—at which the secondary current is lowest.

4468. The greatest density of flux lines in the magnetic circuit of a rotating magnet-type magneto occurs when the magnet is in what position?

1—The neutral position.
2—Full alignment with the field shoe faces.
3—A certain angular displacement beyond the neutral position, referred to as E-gap angle or position.
4—The position where the contact points open.

4469. Magneto breaker point opening relative to the position of the rotating magnet and distributor rotor (internal timing) can be set most accurately

1—after magneto-to-engine timing has been completed.
2—during the magneto-to-engine timing operation, with subsequent in-service readjustment as a result of wear and pitting.
3—during assembly of the magneto before installation on the engine.
4—by setting the points roughly at the required clearance before installing the magneto and then making the fine breaker point adjustment after installation to compensate for wear in the magneto drive train.

4470. Why are high-tension ignition cables frequently routed from the distributors to the spark plugs in flexible metallic conduits?

1—To eliminate high altitude flashover.
2—To reduce the formation of corona and nitric oxide on the cable insulation.
3—To reduce the effect of the high-frequency electromagnetic waves emanated during operation.
4—To decrease the resistance of the current return path (ground).

4471. What will be the results of increasing the gap of the breaker points in a magneto?

1—Retard the spark and increase its intensity.
2—Advance the spark and decrease its intensity.
3—Retard the spark and decrease its intensity.
4—Advance the spark and increase its intensity.

4472. What is the purpose of a safety gap in some magnetos?

1—To discharge the secondary coil's voltage if an open occurs in the secondary circuit.
2—To ground the magneto when the ignition switch is off.
3—To keep the magneto from delivering a spark until it reaches its coming-in speed.
4—To prevent flashover in the distributor.

4473. When timing a magneto internally, the alignment of the timing marks indicates that the

1—breaker points are just closing.
2—magnets are in the neutral position.
3—magnets are in the E–gap position.
4—breaker points are open to their widest gap.

4474. When internally timing a magneto, the breaker points begin to open when the rotating magnet is

1—in the neutral position.
2—fully aligned with the pole shoes.
3—a few degrees past full alignment with the pole shoes.
4—a few degrees past the neutral position.

4475. What is the electrical location of the primary condenser in a high–tension magneto?

1—Across the ignition switch.
2—Across the breaker points.
3—In series with the breaker points.
4—Between the ignition switch and the breaker points.

4476. In a high–tension ignition system, the current in the magneto secondary winding is

1—conducted from the primary winding via the discharge of the condenser.
2—conducted from the primary by the counter e.m.f. developed across the condenser.
3—induced when the primary circuit is interrupted.
4—induced when the primary circuit discharges via the breaker points.

4477. A broken magneto impulse coupling spring will allow which of the following?

1—The magneto on which it is installed becomes inoperative.
2—The magneto to fire one cylinder late in the engine's firing order.
3—The magneto drive gear to turn without moving the magneto rotor.
4—The magneto to operate normally except during start.

4478. What is the radial location of the two north poles of a four–pole rotating magnet in a high–tension magneto?

1—180° apart.
2—270° apart.
3—90° apart.
4—45° apart.

4479. Magneto pole shoes are generally made of

1—laminations of high–grade soft iron.
2—laminations of high–grade alnico.
3—strips of extremely hard steel.
4—pieces of high–carbon iron.

4480. If several long lengths of high–tension ignition cable are to be installed in a rigid shielded ignition manifold, the possibility of damage to the cable as it is pulled through the conduit will be reduced by

1—dusting the cable with powdered soapstone.
2—blowing powdered graphite into the ignition manifold before the cables are installed.
3—the application of a light coat of oil or synthetic grease.
4—dusting the cables with powdered graphite prior to installation.

4481. What component(s) make up the magnetic system of a magneto?

1—Pole shoes, the pole shoe extensions, and the primary coil.
2—Primary and secondary coils.
3—Rotating magnet, the pole shoes, the pole shoe extensions, and the coil core.
4—Rotating magnet.

4482. In an aircraft ignition system, one of the functions of the condenser is to

1—regulate the flow of current between the primary and secondary coil.
2—facilitate a more rapid collapse of the charge in the primary coil.
3—stop the flow of magnetic lines of force when the points open.
4—act as a safety gap for the secondary coil.

4483. When will the voltage in the secondary winding of a magneto, installed on a normally operating engine, be at its highest value?

1—During the power stroke at the point of greatest cylinder pressure.
2—Just prior to spark plug firing.
3—Toward the latter part of the spark duration when the flame front reaches its maximum velocity.
4—Immediately after the breaker points close.

4484. When the switch is off in a battery ignition system, what happens to the primary circuit?

1—A high resistance is connected in series with the primary.
2—The primary circuit is grounded.
3—The primary circuit is opened.
4—The primary circuit is shorted.

4485. As an aircraft engine's speed is increased, the voltage induced in the primary coil of the magneto

1—remains constant.
2—increases.
3—varies with the setting of the voltage regulator.
4—decreases.

4486. When internally timing a magneto, the breaker points begin to open when

1—the piston has just passed TDC at the end of the compression stroke.
2—the flux flow is zero.
3—the magnet poles are a few degrees beyond the neutral position.
4—the magnet poles are fully aligned with the pole shoes.

4487. On a nine–cylinder radial engine, the spark plug wire from the No. 6 distributor block electrode would go to cylinder

1—No. 4.
2—No. 6.
3—No. 8.
4—No. 2.

4488. A safety gap in a magneto is for what purpose?

1—To prevent burning out the primary winding.
2—To protect the high–voltage winding from damage.
3—To prevent arcing across spark plug electrodes.
4—To prevent burning of contact points.

4489. On a seven–cylinder radial engine, No. 3 distributor wire is connected to what cylinder?

1—No. 1.
2—No. 7.
3—No. 5.
4—No. 3.

4490. A defective primary condenser in a magneto is indicated by

1—broken breaker points.
2—a fine–grained frosted appearance of the breaker points.
3—burned and pitted breaker points.
4—a weak spark.

4491. How many secondary coils are required in a low-tension ignition system on an 18–cylinder engine?

1—36.
2—4.
3—18.
4—9.

4492. A magneto ignition switch is connected

1—in series with the breaker points.
2—in series with both the breaker points and the primary condenser.
3—parallel to the breaker points.
4—in series with the primary condenser and parallel to the breaker points.

4493. The spark is produced in a magneto ignition system when the breaker points are

1—beginning to close.
2—fully open.
3—beginning to open.
4—fully closed.

4494. Shielding is used on spark plug and ignition wires to

1—prevent leakage of current which results in a weak spark.
2—protect the wires from short circuits as a result of chafing and rubbing.
3—protect the wires from oil and grease.
4—prevent interference with radio reception.

4495. What is the purpose of using an impulse coupling with a magneto?

1—To absorb impulse vibrations between the magneto and the engine.
2—To compensate for backlash in the magneto and the engine gears.
3—To produce a momentary high rotational speed of the magneto.
4—To prevent the magneto speed from fluctuating at high engine speeds.

4496. When staggered ignition is used, each of the two sparks occurs at a different time. The first spark

1—fires the plug on the intake side of the cylinder.
2—occurs 4° before the second spark.
3—fires the plug on the exhaust side of the cylinder.
4—occurs 4° before E–gap.

4497. The purpose of staggered ignition is to compensate for

1—long ignition harness.
2—short ignition harness.
3—rich fuel/air mixture around exhaust valve.
4—diluted fuel/air mixture around exhaust valve.

4498. Why are most aircraft magneto housings ventilated?

1—To allow excess lubricating oil to drain from the magneto housing.
2—To equalize the pressure inside and outside the magneto housing during flight to prevent the entrance of outside air which may contain moisture.
3—To remove ozone from the magneto housing.
4—To allow heated air from the accessory compartment to keep the internal parts of the magneto dry.

4499. Failure of an engine to cease firing after turning the magneto switch off is an indication of

1—a grounded magneto lead.
2—an open in the low–tension lead to ground.
3—a grounded condenser.
4—a grounded magneto switch.

4500. Alignment of the marks provided for internal timing of a magneto indicates that the

1—breaker points are just beginning to close for No. 1 cylinder.
2—magneto is in E–gap position.
3—No. 1 cylinder is on TDC of compression stroke.
4—distributor gear is correctly aligned to rotor shaft.

4501. When using a timing light to time a magneto to an aircraft engine, the magneto switch should be placed in the

1—BOTH position.
2—OFF position.
3—LEFT position.
4—RIGHT position.

4502. What is the difference between a low–tension and a high–tension engine ignition system?

1—A low–tension system produces relatively low voltage at the spark plug as compared to a high–tension system.
2—A low–tension system does not require any high–voltage ignition leads, but a high–tension system requires all leads to transmit high voltage.
3—A high–tension system is designed for high–altitude aircraft, while a low–tension system is for low– to medium–altitude aircraft.
4—A low–tension system uses a transformer coil near the spark plugs to boost voltage, while the high–tension system voltage is constant from the magneto to the spark plugs.

4503. What test instrument could be used to test a high–tension ignition harness for suspected leakage?

1—A micro–ammeter.
2—A d.c. voltmeter.
3—An a.c. voltmeter.
4—A d.c. ammeter.

4504. If the movable breaker point in a magneto should become stuck in the OPEN position, the magneto

1—will produce an intermittent current at high speed.
2—will fail to produce current.
3—will produce an intermittent current at low speed.
4—secondary winding will become overheated and eventually melt.

4505. The capacitor–type ignition system is used almost universally on turbine engines because of its high voltage and

1—low amperage.
2—long life.
3—low–temperature range.
4—high–heat intensity.

4506. In a low–tension ignition system, each spark plug requires an individual

1—condenser.
2—cam assembly.
3—breaker assembly.
4—secondary coil.

4507. A certain nine–cylinder radial engine used a noncompensated single–unit, dual–type magneto with a four–pole rotating magnet and separately mounted distributors. Which of the following will have the lowest RPM at any given engine speed?

1—The breaker cam.
2—The engine crankshaft.
3—The distributors.
4—The rotating magnet.

4508. The No. 8 ignition lead from the left magneto distributor on a 14–cylinder radial engine is connected to the

1—front spark plug of No. 11 cylinder.
2—front spark plug of No. 5 cylinder.
3—rear spark plug of No. 8 cylinder.
4—rear spark plug of No. 2 cylinder.

4509. What will be the effect if the spark plugs are gapped too wide?

1—Insulation failure.
2—Hard starting.
3—Shell breakdown.
4—Lead damage.

4510. When removing a shielded spark plug, which of the following is most likely to be damaged?

1—Center electrode.
2—Shell section.
3—Ground electrodes.
4—Core insulator.

4511. What effect would a cracked distributor rotor have on a magneto?

1—Ground the secondary circuit through the crack.
2—Fire the trailing cylinder.
3—Fire two cylinders simultaneously.
4—Ground the primary circuit through the crack.

4512. How does the ignition system of a gas–turbine engine differ from that of a reciprocating engine?

1—Magneto–to–engine timing is not critical.
2—One spark plug is used in each combustion chamber.
3—A high–voltage, high–energy spark is required for ignition.
4—Low–energy igniter plugs are used in place of spark plugs.

4513. In a turbojet engine d.c. capacitor discharge ignition system, where are the high–voltage pulses formed?

1—At the breaker.
2—At the triggering transformer.
3—At the rectifier.
4—At the multilobe cam.

4514. Which of the following breaker point characteristics is associated with a faulty condenser?

1—Oily.
2—Crowned.
3—Fine grained.
4—Coarse grained.

4515. How are most radial engine spark plug wires connected to the distributor block?

1—By use of cable–piercing screws.
2—By use of self–locking cable ferrules.
3—By use of terminal sleeves and retaining nuts.
4—By friction between the cable ferrule and distributor block well.

4516. Thermocouples are usually inserted or installed on the

1—coldest cylinder of the engine.
2—front cylinder of the engine.
3—rear cylinder of the engine.
4—hottest cylinder of the engine.

4517. In a high voltage d.c. capacitor input turbine engine ignition system, the contact points for the ignition system are closed by

1—a magnetic force.
2—spring action.
3—cam action.
4—high capacitance discharge.

4518. A radio noise filter in a turbine engine ignition system filters radio frequency noise pulses by

1—reducing the voltage spikes.
2—increasing the noise frequencies to a level that is compatible to the radio equipment.
3—blocking radio frequency noise pulses and shunting them to ground.
4—inducting voltage spikes.

4519. A test of a reciprocating engine ignition harness revealed excessive leakage in a majority of the leads. What is a probable cause?

1—An improper ground of lead shielding.
2—A shorted primary coil in the magneto.
3—A deteriorated condition of the distributor block.
4—Improper spark plug gap.

4520. Why are turbine engine igniters less susceptible to fouling than reciprocating engine spark plugs?

1—The high–intensity spark cleans the igniter with heat.
2—The frequency of the spark is less for igniters.
3—Turbine igniters operate cooler.
4—Turbine fuel does not contain igniter contaminates.

4521. The constrained–gap igniter plug used in some gas turbine engines operates at a cooler temperature because

1—it projects into the combustion chamber.
2—the applied voltage is less.
3—the construction is such that the spark occurs beyond the face of the combustion chamber liner.
4—it has multiple electrodes to share the voltage arcing.

4522. What should be used to clean grease or carbon tracks from condensers or coils that are used in magnetos?

1—Solvent.
2—Acetone.
3—Soap and water.
4—Naptha.

4523. What controls the spark rate of the igniter plug on a turbine engine with a capacitor–type ignition system?

1—D.C. voltage applied directly to the igniter plug.
2—RPM of the starting motor.
3—RPM of the d.c. motor in the ignition unit.
4—Resistance in the primary ignition lead.

4524. The igniters used in turbine engines have a wider gap than spark plugs used in reciprocating engines because

1—the applied voltage is much greater.
2—the electrode material has less resistance.
3—operating pressures under which the igniter fires is much less.
4—the applied voltage is much less.

4525. Igniter plugs used in turbine engines are subjected to much higher voltage than reciprocating engine spark plugs, yet the service life is longer. This is because

1—they operate at much lower temperatures.
2—the electrode gap is much smaller.
3—they are not placed directly into the combustion area.
4—they do not require continuous operation.

4526. Which of the following will cause the center electrode insulator of ceramic spark plugs to fracture and/or break?

1—Improper timing.
2—Electrical erosion.
3—Improper gapping procedures.
4—Excessive magneto voltage.

4527. When using the ignition analyzer, what determines the height of the pattern line above the center of the screen?

1—Primary voltage in the ignition system.
2—Primary current at the magneto breaker points.
3—Secondary current at the high–tension lead.
4—Secondary voltage at the spark plug.

4528. The sparking order of a distributor used on a nine–cylinder radial engine is

1—1, 3, 5, 7, 9, 2, 4, 6, and 8.
2—1, 3, 2, 5, 7, 4, 6, 9, and 8.
3—1, 2, 4, 6, 8, 3, 5, 9, and 7.
4—1, 2, 3, 4, 5, 6, 7, 8, and 9.

4529. In a high–tension ignition system, a primary condenser of too low a capacity will cause

1—excessive primary voltage.
2—excessively high secondary voltage.
3—the breaker contacts to burn.
4—excessive burning of the spark plug electrodes.

4530. Which of the following, obtained during magneto check at 1700 RPM, indicates a short (grounded) circuit between the right magneto primary and the ignition switch?

1—BOTH—1700 RPM,
  R—1625 RPM,
  L—1700 RPM,
  OFF—1625 RPM.
2—BOTH—1700 RPM,
  R—0 RPM,
  L—1700 RPM,
  OFF—0 RPM.
3—BOTH—1700 RPM,
  R—1625 RPM,
  L—1675 RPM,
  OFF—1625 RPM.
4—BOTH—1700 RPM,
  R—0 RPM,
  L—1675 RPM,
  OFF—0 RPM.

4531. If an aircraft ignition switch is turned off and the engine continues to run normally, the trouble is probably caused by

1—a leak in the carburetor.
2—an open ground lead in the magneto.
3—arcing magneto breaker points.
4—failure to turn off the battery switch.

4532. Where are the breaker points located in a typical low–tension ignition system of an 18–cylinder radial engine used on a certificated aircraft?

1—In the magneto, and each set of points is associated with a transformer.
2—In the distributors, and are actuated by four compensating cams.
3—In the magneto, and are actuated by one 18–lobe compensating cam.
4—In the distributor, and each set of points is associated with 18 transformers.

4533. Which statement is correct regarding the ignition system of a turbojet engine?

1—The system is normally de–energized as soon as the engine starts.
2—It is a low–voltage, high–amperage system.
3—It is energized during the starting and warmup periods only.
4—The system generally includes a polar inductor–type magneto.

4534. When the ignition switch of a single–engine aircraft is turned to the OFF position,

1—the primary circuits of both magnetos are grounded.
2—the secondary circuits of both magnetos are opened.
3—all circuits are automatically opened.
4—the high–tension lead from the battery is grounded.

4535. Which cylinder is being fired if the distributor finger of a 14–cylinder radial engine is pointing to No. 7 electrode?

1—No. 10.
2—No. 6.
3—No. 7.
4—No. 13.

4536. On a double–row radial engine when the ignition switch is on LEFT,

1—all plugs in the front row are firing.
2—all plugs in the rear row are firing.
3—the front plugs of both rows are firing.
4—the rear plugs of both rows are firing.

4537. How are the operating temperatures of a spark plug controlled?

1—Area of the plug exposed to the cooling airstream.
2—Rate of heat transfer of the engine seat gasket.
3—Area of the plug terminal.
4—Rate of heat inductance of the plug.

4538. What will be the result if the secondary winding in a low–tension ignition coil fails?

1—One spark plug will fail to fire.
2—All plugs will fail to fire.
3—One row of plugs (front or rear) will fail to fire.
4—One set (both plugs in one cylinder) will fail to fire.

4539. If staggered ignition timing is used, the

1—spark plug nearest the exhaust valve will fire first.
2—spark will be automatically retarded as engine speed increases.
3—spark will be automatically advanced as engine speed increases.
4—spark plug nearest the intake valve will fire first.

4540. The term reach, as applied to spark plug design and/or type, indicates

1—the length of the center electrode insulation exposed to the flame of combustion.
2—the linear distance from the shell gasket seat to the end of the shell skirt.
3—the length of center electrode exposed to the flame of combustion.
4—the length of the shielded barrel.

4541. When inspecting the ignition cable installation on a nine–cylinder radial engine, the spark plug lead coming from the No. 5 opening in the right magneto distributor block should go to which plug and cylinder?

1—Rear plug of No. 9 cylinder.
2—Front plug of No. 9 cylinder.
3—Front plug of No. 5 cylinder.
4—Rear plug of No. 5 cylinder.

4542. The numbers appearing on the ignition distributor block indicate the

1—sparking order of the distributor.
2—relation between distributor terminal numbers and cylinder numbers.
3—ratio of the distributor rotor speed to the crankshaft speed.
4—firing order of the engine.

4543. Which statement is correct regarding the ignition system of a turbojet engine?

1—The system is operated during starting and warmup only.
2—The system is de–energized after the engine starts.
3—It is a low–energy system designed to operate in a high–temperature environment.
4—It is a low–voltage, high–amperage system.

4544. When testing a magneto distributor block for electrical leakage, which of the following pieces of test equipment should be used?

1—A high–tension harness tester.
2—A condenser tester.
3—A continuity tester.
4—A high–range ammeter.

4545. A careful examination of a set of ceramic spark plugs disclosed several fractured and broken insulator core tips. This type of insulation damage is generally a result of

1—service vibration.
2—improper gapping procedures.
3—thermal shocks.
4—excessive magneto voltage.

4546. Compensated timing provides for the firing of the cylinders

1—at the position of crankshaft travel that provides the balance between the inertia of the reciprocating mass and the force that results from compresssing the fuel/air charge.
2—at the piston position which produces peak compression regardless of the degree of crankshaft travel required to obtain the position.
3—an equal number of degrees of crankshaft travel apart regardless of variations in piston position caused by articulation of the connecting rod assembly.
4—in relationship to piston position regardless of variations in crankshaft travel required to obtain the position.

4547. Which of the following does not relate to the heat characteristics of a spark plug?

1—Spark plug reach.
2—Area of the spark plug exposed to the cooling air.
3—Area of the center electrode exposed to the heat of combustion.
4—Material from which the spark plug is constructed.

4548. Hot spark plugs are generally used in aircraft powerplants

1—with comparatively high compression or high operating temperatures.
2—with comparatively low operating temperatures.
3—which are loosely baffled.
4—which produce high power per cubic inch displacement.

4549. If a spark plug lead becomes grounded, the

1—the magneto secondary winding will become overloaded and break down.
2—magneto will not be affected.
3—distributor rotor finger will discharge to the next closest electrode within the distributor.
4—condenser will break down.

4550. Which of the following statements regarding magneto switch circuits is not true?

1—In the BOTH position, the right and left magneto circuits are grounded.
2—In the OFF position, neither the right nor left magneto circuits are open.
3—In the RIGHT position, the right magneto circuit is open and the left magneto circuit is grounded.
4—In the LEFT position, the left magneto circuit is open and the right magneto circuit is grounded.

4551. Which of the following statements most accurately describes spark plug heat range?

1—The length of the threaded portion of the shell usually denotes the spark plug heat range.
2—A hot plug is designed so that the insulator tip is reasonably short to hasten the rate of heat transfer from the tip through the spark plug shell to the cylinder head.
3—A cold plug is designed so that the insulator tip is reasonably short to hasten the rate of heat transfer from the tip through the spark plug shell to the cylinder head.
4—Shielded spark plugs have a much higher heat range than nonshielded spark plugs.

4552. When does battery current flow through the primary circuit of a battery ignition coil?

1—Only when a spark plug is firing.
2—Only when the breaker points are open.
3—At all times that the ignition switch is on.
4—When the breaker points are closed and the ignition switch is on.

4553. In order to turn a magneto off, the primary circuit must be

1—shunted to the battery circuit.
2—grounded.
3—opened.
4—shorted.

4554. When performing a magneto ground check on an engine, correct operation is indicated by

1—a decrease in manifold pressure.
2—an increase in RPM.
3—no drop in RPM.
4—a slight drop in RPM.

4555. Defective spark plugs will cause

1—intermittent missing of the engine at high speeds only.
2—intermittent missing of the engine at low speeds only.
3—failure of the magneto.
4—intermittent missing of the engine at all speeds.

4556. A spark plug is fouled when

1—its gap is too small.
2—its magneto wire is not connected.
3—it causes preignition.
4—its spark grounds without jumping electrodes.

4557. Which of the following would be cause for rejection of a spark plug?

1—Carbon fouling of the electrode and insulator.
2—Insulator tip cracked.
3—Center electrode being a light grey color.
4—Lead fouling of the electrode and insulator.

4558. What will be the result of using too hot a spark plug?

1—Failure of the engine.
2—Fouling of plug.
3—Preignition.
4—Burned condenser.

4559. Upon inspection of the spark plugs in an aircraft engine, the plugs were found caked with a heavy black soot. This indicates

1—worn oil seal rings.
2—a rich mixture.
3—a lean mixture.
4—improper spark plug gap setting.

4560. Spark plug heat range is determined by

1—the reach of the spark plug.
2—its ability to transfer heat to the cylinder head.
3—the number of ground electrodes.
4—outside air temperature.

4561. Ignition check during engine runup indicates excessive RPM drop during operation on right magneto. The major portion of the RPM loss occurs rapidly after switching to the right magneto position (fast drop). The most likely cause is

1—faulty or fouled spark plugs.
2—incorrect ignition timing on both magnetos.
3—incorrect valve timing.
4—one or more dead cylinders.

4562. If new breaker points are installed in a magneto on an engine, it will be necessary to time the

1—magneto internally and the magneto to the engine.
2—breaker points to the No. 1 cylinder.
3—magneto drive to the engine.
4—distributor gear to the magneto drive.

4563. Using a cold spark plug in a high–compression aircraft engine would probably result in

1—normal operation.
2—preignition.
3—a fouled plug.
4—detonation.

4564. Spark plug fouling caused by lead deposits

1—occurs most often during cruise with rich mixture.
2—occurs most often when cylinder head temperatures are relatively low.
3—may best be corrected by increasing engine RPM to burn away the lead deposits.
4—occurs most often when cylinder head temperatures are high.

4565. In a four-stroke cycle aircraft engine, when does the ignition event take place?

1—After the piston reaches TDC on intake stroke.
2—Before the piston reaches TDC on compression stroke.
3—After the piston reaches TDC on power stroke.
4—After the piston reaches TDC on compression stroke.

4566. Which of the following conditions occur when timing an engine ignition system?

1—The rotating magnet of the magneto must not be in the E–gap position.
2—The piston in the No. 1 cylinder must be a prescribed number of degrees before top center on the compression stroke.
3—The magneto breaker points must be closed.
4—The magneto breaker points must be in the E–gap position.

4567. The spark occurs at the spark plug when the ignition's

1—secondary circuit is broken.
2—secondary circuit is completed.
3—primary circuit is completed.
4—primary circuit is broken.

4568. The type of ignition system used on most jet aircraft engines is

1—high resistance.
2—magneto.
3—low tension.
4—capacitor discharge.

4569. Ignition check during engine runup indicates a slow drop in RPM. This is usually caused by

1—defective spark plugs.
2—a defective high–tension lead.
3—incorrect ignition timing or valve adjustment.
4—distributor points too wide.

4570. If the ground wire of a magneto is disconnected at the ignition switch, the result will be

1—the affected magneto will be isolated and the engine will run on the opposite magneto.
2—a decrease in magnetic lines of force.
3—the engine will stop running.
4—the engine will not stop running when the ignition switch is turned off.

4571. Which of the following are advantages of dual ignition in aircraft engines?

A. Gives a more complete and quick combustion of the fuel.
B. Provides a backup magneto system.
C. Increases the output power of the engine.
D. Permits the use of lower grade fuels.
E. Increases the intensity of the spark at the spark plugs.

1—B, C, D.
2—C, D, E.
3—B, C, E.
4—A, B, C.

4572. How does high–tension ignition shielding tend to reduce radio interference?

1—Prevents ignition flashover at high altitudes.
2—Protects the ignition leads against the entrance of moisture and subsequent electrical leakage.
3—Reduces voltage drop in the transmission of high–tension current.
4—Receives and grounds high–frequency waves coming from the magneto and high–tension ignition leads.

4573. Which of the following are distinct circuits of a high–tension magneto?

A. Magnetic.
B. Primary.
C. E–gap.
D. P–lead.
E. Secondary.

1—A, B, E.
2—A, C, D.
3—B, E, D.
4—B, C, D.

4574. What are two parts of a distributor in an aircraft engine ignition system?

A. Coil.
B. Block.
C. Stator.
D. Rotor.
E. Transformer.

1—B and D.
2—A and C.
3—C and D.
4—B and E.

4575. What is a result of "flashover" in a distributor?

1—An intense voltage at the spark plug.
2—A reversal of current flow.
3—Erosion of the cigarette.
4—A conductive carbon trail.

4576. What is the relationship between distributor and crankshaft speed of aircraft reciprocating engines?

1—The distributor turns at crankshaft speed.
2—The distributor turns at one–half crankshaft speed.
3—The distributor turns one and one–half crankshaft speed.
4—The crankshaft turns at one–half distributor speed.

4577. Why do turbine engine ignition systems require high energy?

1—To ignite the fuel under conditions of high altitude and high temperatures.
2—Because the applied voltage is much greater.
3—To ignite the fuel under conditions of high altitude and low temperatures.
4—Because the applied voltage is much less.

4578. Which of the following are included in a typical turbine engine ignition system?

A. Two exciter units.
B. One exciter unit.
C. Two transformers.
D. Two intermediate ignition leads.
E. Two low–tension leads.
F. Two high–tension leads.
G. One transformer.

1—B, C, D, E.
2—A, D, E, G.
3—A, D, F, G.
4—A, C, D, F.

4579. At what RPM is a reciprocating engine ignition switch check made?

1—1500 RPM.
2—The slowest possible RPM.
3—Full throttle RPM.
4—Cruise RPM.

4580. What is the position of the rotating magnet in a high–tension magneto when the breaker points are closed?

1—Neutral.
2—Full register.
3—E–gap.
4—A few degrees after neutral.

4581. What component of a dual magneto is shared by both ignition systems?

1—Breaker points.
2—High–tension coil.
3—Rotating magnet.
4—Capacitor.

4582. What would be the result if a magneto breaker point mainspring did not have sufficient tension?

1—The points will stick.
2—The points will not open to the specified gap.
3—The points will float or bounce.
4—A greater than normal induction buildup of the magneto.

4583. The secondary coil of a magneto is grounded through the

1—ignition switch.
2—grounded side of the primary condenser.
3—primary coil.
4—grounded side of the breaker points.

4584. What will occur if the water injection switch is turned on for a check of the carburetor derichment valve with the engine inoperative?

1—The valve will close because of water pump pressure.
2—The valve will not actuate because of the absence of oil pressure.
3—The valve will open by means of an electric solenoid mechanism.
4—The valve will close because of fuel boost pump pressure.

4585. A carburetor is equipped with a derichment valve and a derichment jet which adds a cooling fluid. This is called

1—water evaporator additive.
2—an injection of water and freon.
3—atmosphere injection.
4—ADI (anti–detonant injection).

4586. When the water injection system on a turbine engine airplane contains water and is armed in the cockpit,

1—the water is turned on automatically when a predetermined EGT is reached.
2—the water injection system is turned on by a timer actuated by the power lever.
3—the water injection valves are opened by a switch on their respective power levers in the cockpit.
4—nothing happens until the outside air temperature exceeds 100° F.

4587. The anti–detonant fluid used in water injection systems is a mixture of

1—water and benzine.
2—alcohol and water.
3—potassium dichromate and water.
4—none of the above.

4588. If the water injection switch on a reciprocating engine is turned on when the engine is not operating, the derichment valve will

1—close because of water pressure.
2—open by means of an electric solenoid.
3—not actuate.
4—close if fuel pressure is available.

4589. In addition to permitting an increase in maximum manifold pressure, the water injection system permits the engine to

1—operate at METO power.
2—operate at maximum power for unlimited periods.
3—develop increased power without changing the manifold pressure and RPM settings.
4—increase the aircraft cruising range by replacing part of the fuel/air mixture with an anti-detonant.

4590. How does injection of water or water-alcohol during high-power output increase the available power of reciprocating engines?

1—By increasing the weight of charge.
2—By suppressing detonation.
3—By improving volumetric efficiency.
4—By increasing the burning rate of the fuel/air charge.

4591. What actuates the derichment valve in a pressure carburetor?

1—Throttle linkage.
2—Air pressure.
3—Water (ADI) pressure.
4—Fuel pressure.

4592. The automatic fuel-flow metering mechanisms of most modern carburetors are actuated by the

1—velocity of the air passing through the carburetor.
2—velocity as well as the mass of air passing through the carburetor.
3—mass of air passing through the carburetor.
4—position of the throttle.

4593. On a float-type carburetor, the purpose of the economizer valve is to

1—economize on the amount of fuel discharged into the induction system.
2—provide extra fuel for sudden acceleration of the engine.
3—maintain the leanest mixture possible during cruising best power.
4—provide a richer mixture and fuel cooling at maximum power output.

4594. The fuel metering force of a conventional float-type carburetor in its normal operating range is the difference between the pressure acting on the discharge nozzle located within the venturi and the pressure

1—acting on the fuel in the float chamber.
2—of the fuel as it enters the carburetor.
3—of the air as it enters the venturi (impact pressure).
4—on the downstream or engine side of the throttle valve.

4595. If the main air bleed of a float-type carburetor becomes clogged, the engine will run

1—lean at rated power.
2—rich at rated power.
3—rich at idling.
4—lean at idling.

4596. Which method is commonly used to adjust the level of a float in a float-type carburetor?

1—Lengthening or shortening the float shaft.
2—Add or remove shims under the needle-valve seat.
3—Change the angle of the float arm pivot.
4—Add or remove float weights.

4597. As the density of air decreases with increased altitude, the AMC (Automatic Mixture Control) unit on a pressure carburetor will cause the air metering force to

1—increase by restricting the flow of air from chamber B (boost venturi suction).
2—decrease by restricting the flow of air to chamber A (impact pressure).
3—increase by reducing the restriction to the flow of air to chamber A (impact pressure).
4—decrease by reducing the restriction to the flow of air from chamber B (boost venturi suction).

4598. What is the possible cause of an engine running rich at full throttle if it is equipped with a float-type carburetor?

1—Float level too low.
2—Clogged main air bleed.
3—Clogged atmospheric vent.
4—Restricted main metering fuel jet.

4599. Why does a float-type carburetor with a plugged main air bleed run rich?

1—Pressure in the float chamber has been decreased.
2—More of the available suction will act on the fuel in the discharge nozzle.
3—Less of the available suction will act on the fuel in the discharge nozzle.
4—Pressure in the float chamber has been increased.

4600. A punctured float in a float–type carburetor will cause the fuel level to

1—lower, and enrich the mixture.
2—rise, and enrich the mixture.
3—rise, and lean the mixture.
4—lower, and lean the mixture.

4601. Upon what principle does the back–suction type mixture control operate?

1—By varying the pressure within the venturi section.
2—By altering the height of the fuel in the float chamber.
3—By varying the pressure acting on the fuel in the float chamber.
4—By changing the effective cross–sectional area of the main metering orifice (jet).

4602. If an aircraft engine is equipped with a carburetor that is not compensated for altitude and temperature variations, the fuel/air mixture will become

1—leaner as either the altitude or temperature increases.
2—richer as the altitude increases and leaner as the temperature increases.
3—richer as either the altitude or temperature increases.
4—leaner as the altitude increases and richer as the temperature increases.

4603. Float–type carburetors which are equipped with economizers are normally set for

1—economizer valves to be open at cruising speeds and closed at maximum RPM.
2—their leanest mixture delivery and enriched by means of the economizer system.
3—their richest mixture delivery and leaned by means of the economizer system.
4—the economizer system to supplement the main system supply at all engine speeds above idling.

4604. If a float–type carburetor becomes flooded, the condition is most likely caused by

1—a leaking needle valve and seat assembly.
2—a clogged main discharge nozzle.
3—the accelerating pump shaft being stuck.
4—a clogged back–suction line.

4605. If an engine is equipped with a float–type carburetor and the engine runs rich at full throttle, a possible cause of the trouble is a clogged

1—main air bleed.
2—back–suction line.
3—atmospheric vent line.
4—main metering fuel jet.

4606. What occurs when a back–suction type mixture control is placed in IDLE CUTOFF?

1—The fuel passages to the main and idle jets will be closed by a valve.
2—The float chamber will be vented to a negative pressure area.
3—The fuel passage to the idle jet will be closed by a valve.
4—The fuel passage to the main jet will be closed by a valve.

4607. Which of the following best describes the function of an altitude mixture control?

1—Regulates the richness of the fuel/air charge entering the engine.
2—Regulates the air pressure above the fuel in the float chamber.
3—Regulates the air pressure in the venturi.
4—Regulates the main airflow to the engine.

4608. Select the correct statement concerning the idle system of a conventional float–type carburetor.

1—The low–pressure area created in the throat of the venturi pulls the fuel from the idle passage.
2—Climatic conditions have very little effect on idle mixture requirements.
3—The low pressure between the edges of the throttle valve and the throttle body pulls the fuel from the idle passage.
4—Airport altitude has very little effect on idle mixture requirements.

4609. On an engine equipped with a pressure–type carburetor, fuel supply in the idling range is insured by the inclusion in the carburetor of

1—a separate fuel supply to the discharge nozzle that supplements normal carburetor fuel supply in the idle range.
2—a spring in the unmetered fuel chamber to supplement the action of normal metering forces.
3—an idle metering jet that bypasses the carburetor in the idle range.
4—a separate boost venturi that is sensitive to the reduced airflow at start and idle speeds.

4610. The economizer system of a float–type carburetor performs which of the following functions?

1—It supplies and regulates the fuel required for all engine speeds below cruising.
2—It supplies and regulates the fuel required for all engine speeds.
3—It supplies and regulates the additional fuel required for all engine speeds above cruising.
4—It regulates the fuel required for all engine speeds and all altitudes.

4611. How will the mixture of an engine be affected if the bellows of the AMC (automatic mixture control) in a pressure carburetor ruptures while the engine is operating at altitude?

1—It will become leaner.
2—No change will occur until the altitude changes.
3—No change will occur until the throttle setting is changed.
4—It will become richer.

4612. The fuel level within the float chamber of a properly adjusted float–type carburetor will be

1—slightly higher than the discharge nozzle outlet.
2—unrelated to the discharge nozzle outlet position.
3—slightly lower than the discharge nozzle outlet.
4—at the same level as the discharge nozzle outlet.

4613. How would the mixture of an engine be affected if the diaphragm between chambers A (regulated air inlet pressure) and B (boost venturi pressure) ruptured on a pressure injection carburetor?

1—It will become rich at all power settings.
2—It will become rich at low power settings.
3—It will become lean at all power settings.
4—It will become lean at low power settings.

4614. Rupture or leakage of the air diaphragm between chambers A (regulated air inlet pressure) and B (boost venturi pressure) of an injection carburetor will cause the fuel/air mixture to become

1—lean at low power settings and rich at all power settings which utilize the power enrichment system.
2—lean at all engine power settings and operating altitudes.
3—lean on unsupercharged engines and rich on supercharged engines.
4—rich at all engine power settings and operating altitudes.

4615. The metered fuel pressure (chamber C) in an injection–type carburetor

1—is held constant throughout the entire engine operating range.
2—varies according to the position of the poppet valve located between chamber D (unmetered fuel) and chamber E (engine–driven fuel pump pressure).
3—will be approximately equal to the pressure in chamber A (impact pressure).
4—varies in proportion to the mass airflow being handled by the carburetor.

4616. As the throttle of an engine equipped with a pressure injection carburetor is retarded, the pressure differential between air chambers A (regulated air inlet pressure) and B (boost venturi pressure) will

1—remain constant if the throttle is closed slowly.
2—decrease.
3—decrease momentarily, then increase to previous differential pressure.
4—increase momentarily, then decrease to previously differential pressure.

4617. Select the statement which is correct relating to a fuel level check of a float–type carburetor.

1—Do not use leaded gasoline.
2—Use 5 pounds fuel pressure for the test if the carburetor is to be used in a gravity fuel feed system.
3—Block off the main and idle jets to prevent a continuous flow of fuel through the jets.
4—Do not measure the level at the edge of the float chamber.

4618. Which statement regarding the air metering force in a pressure–injection carburetor installed on a 14–cylinder engine is correct?

1—The air metering force opens or closes the poppet valve according to the pressure drop in the boost venturi.
2—The air metering force is balanced solely by the fuel metering force during all operation from idling to full power.
3—The air metering force is modified by the operation of the manual mixture control.
4—The air metering force is assisted by the idle spring which holds the poppet valve open during low RPM operation.

4619. What carburetor component measures the amount of air delivered to the engine?

1—Economizer valve.
2—Automatic mixture control.
3—Cloverleaf.
4—Venturi.

4620. The automatic mixture control on a pressure carburetor controls the fuel/air ratio by restricting the

1—impact air passage to chamber A.
2—fuel flow at the discharge nozzle.
3—boost venturi pressure in chamber B.
4—metered fuel in chamber C.

4621. When starting an aircraft engine equipped with carburetor heat, the heat control should be placed in

1—the cold position.
2—the hot position.
3—between the hot and cold position.
4—the warm filtered air position.

4622. Fuel is discharged for idling speeds on a float–type carburetor

1—through the main discharge nozzle.
2—from the idle discharge nozzle.
3—in the venturi.
4—through the idle discharge air bleed.

4623. When air passes through the venturi of a carburetor, what three changes occur?

1—Velocity increases, temperature increases, and pressure decreases.
2—Velocity decreases, temperature decreases, and pressure decreases.
3—Velocity decreases, temperature increases, and pressure increases.
4—Velocity increases, temperature decreases, and pressure decreases.

4624. Where is the throttle valve located on a float–type carburetor?

1—Between the venturi and the discharge nozzle.
2—After the main discharge nozzle and venturi.
3—Before the venturi, but after the butterfly valve.
4—After the venturi and just before the main discharge nozzle.

4625. The parting surface of a float–type carburetor is used to

1—determine the main discharge nozzle height.
2—adjust the height of the economizer needle.
3—measure the level of the accelerating pump piston.
4—determine the float level.

4626. An aircraft carburetor is equipped with a mixture control in order to prevent

1—ice formation in the carburetor.
2—the mixture from becoming too lean at high altitudes.
3—the mixture from becoming too rich at high altitudes.
4—the mixture from becoming too rich at high speeds.

4627. Which of the following is not a function of the carburetor venturi?

1—Proportions the air/fuel mixture.
2—Decreases pressure at the discharge nozzle.
3—Regulates the idle system.
4—Limits the airflow at full throttle.

4628. Idle cutoff is accomplished on a carburetor equipped with a back–suction mixture control by

1—introducing low pressure (intake manifold) air into the float chamber.
2—manually raising the float.
3—turning the fuel selector valve to OFF.
4—the positive closing of a needle and seat.

4629. The primary purpose of an air bleed in a float–type carburetor is to

1—aid fuel vaporization and control fuel discharge.
2—meter air to adjust the mixture.
3—decrease fuel density and destroy surface tension.
4—vent the back–suction mixture control.

4630. To determine the float level in a float–type carburetor, a measurement is usually made from the top of the fuel in the float chamber to the

1—parting surface of the carburetor.
2—top of the float.
3—bottom of the float chamber.
4—centerline of the main discharge nozzle.

4631. The throttle valve of float–type aircraft carburetors is located

1—ahead of the venturi and main discharge nozzle.
2—after the main discharge nozzle and ahead of the venturi.
3—between the venturi and the engine.
4—after the venturi and ahead of the main discharge nozzle.

4632. Which statement relative to the AMC (automatic mixture control) is true?

1—The AMC is placed in series with the venturi.
2—The AMC is placed in parallel with the boost venturi.
3—The AMC compensates for changes in air density thereby enabling the proper amount of fuel to be delivered.
4—An increase in air density results in a decrease of airflow through the boost venturi.

4633. Why must a float–type carburetor supply a rich mixture during idle?

1—Because of poor cylinder scavenging during idle.
2—The spark available for ignition during low RPM will not ignite a normal mixture.
3—Because of the effect of accumulative intake pipe leaks at idle.
4—Engine operation at low RPM results in higher than normal volumetric efficiency.

4634. What component is used to ensure fuel delivery during periods of rapid engine acceleration?

1—Acceleration pump.
2—Standby carburetor.
3—Water injection pump.
4—Power enrichment unit.

4635. The device that controls the ratio of the fuel/air mixture to the cylinders is called

1—a throttle valve.
2—a mixture control.
3—an acceleration pump.
4—a metering jet.

4636. The device that controls the volume of the fuel/air mixture to the cylinders is called

1—an acceleration pump.
2—a mixture control.
3—a metering jet.
4—a throttle valve.

4637. Which statement is correct regarding a continuous–flow fuel injection system used on some reciprocating engines?

1—Fuel is injected directly into each cylinder.
2—Fuel is injected at each cylinder intake port.
3—The injection system must be timed to the engine.
4—Two injector nozzles are used in the injector fuel system for various speeds.

4638. During the operation of an aircraft engine, the pressure drop in the carburetor venturi depends primarily upon the

1—air temperature.
2—barometric pressure.
3—air velocity.
4—humidity.

4639. The double–diaphragm acceleration pump used with an injection carburetor is discharged by

1—a sudden drop in pressure between the throttle valve and the supercharger impeller.
2—suitable linkage connected to the throttle shaft.
3—a sudden rise in pressure between the throttle valve and the supercharger impeller.
4—suitable linkage connected to the fuel–metering head enrichment valve.

4640. Which of the following causes a single diaphragm accelerator pump to discharge fuel?

1—An increase in venturi suction when the throttle valve is open.
2—An increase in manifold pressure that occurs when the throttle valve is opened.
3—A decrease in manifold pressure that occurs when the throttle valve is opened.
4—A decrease in pressure differential acting on the fuel metering orifices (jets) when the throttle valve is opened.

4641. At what engine speed does the main metering jet actually function as a metering jet in a float–type carburetor?

1—All RPM's.
2—Crusing RPM only.
3—Maximum RPM only.
4—All RPM's above idle range.

4642. An aircraft engine continuous cylinder fuel injection system normally discharges fuel into each cylinder head intake valve port during which stroke(s)?

1—Intake.
2—Compression.
3—Intake and compression.
4—All (continuously).

4643. What is the purpose of the carburetor accelerating system?

1—To supply and regulate the fuel required for engine speeds above idle.
2—Temporarily enrich the mixture when the throttle is suddenly opened.
3—To supply and regulate additional fuel required for engine speeds above cruising.
4—Temporarily derich the mixture when the throttle is suddenly closed.

4644. When changing a float–type carburetor removed from a 150–hp. engine for use with a 180–hp. engine, which of the following components would probably require change?

1—Venturi.
2—Float.
3—Throttle valve.
4—Needle–valve and seat assembly.

4645. A fuel injection system used on reciprocating engines where fuel is injected at the intake valve port

1—is referred to as a continuous–flow system.
2—employs a fuel pump which is timed to the engine.
3—requires a fuel metering unit to atomize the fuel.
4—employs two pumps which are synchronized with a bar.

4646. What is the relationship between the accelerating pump and the enrichment valve in a pressure–injection carburetor?

1—No relationship since they operate independently.
2—Fuel pressure affects both units.
3—The accelerating pump actuates the enrichment valve.
4—The mixture control changes the setting of each unit.

4647. What is the relationship between the pressure existing within the throat of a venturi and the velocity of the air passing through the venturi?

1—There is no direct relationship between the pressure and the velocity.
2—The pressure is proportional to the square of the velocity.
3—The pressure is directly proportional to the velocity.
4—The pressure is inversely proportional to the velocity.

4648. Which of the following is least likely to occur during operation of an engine equipped with a direct cylinder fuel injection system?

1—Torching.
2—Afterfiring.
3—Kickback during start.
4—Backfiring.

4649. During the operation of a reciprocating engine, the pressure drop in the carburetor venturi depends primarily upon the

1—air temperature.
2—fuel/air mixture.
3—barometric pressure.
4—air velocity.

4650. What carburetor component actually limits the desired maximum airflow to the engine at full throttle?

1—Throttle valve.
2—Venturi.
3—Manifold intake.
4—Air diaphragm.

4651. On a carburetor without an automatic mixture control as you ascend to altitude, the mixture will

1—be enriched.
2—be leaned.
3—remain at the same ratio.
4—not be affected.

4652. During engine operation, if carburetor heat is applied, it will

1—increase air to fuel ratio.
2—decrease carburetor air temperature.
3—increase engine RPM.
4—decrease the air volume through the carburetor.

4653. The desired engine idle speed and mixture setting

1—is adjusted with engine warmed up and operating.
2—should give minimum RPM with maximum manifold pressure.
3—is usually adjusted in the following sequence; speed first, then mixture.
4—is adjusted with the throttle advanced.

4654. A nine-cylinder radial engine, using a multiple-point priming system with a central spider, will prime which cylinders?

1—One, two, three, eight, and nine.
2—All cylinders.
3—Top three cylinders.
4—One, three, five, and seven.

4655. What is a function of the idling air bleed in a float-type carburetor?

1—It provides a means for adjusting the mixture at idle speeds.
2—It vaporizes the fuel at idling speeds.
3—It provides a means for adjusting the idle speed.
4—It aids in emulsifying the fuel at idle speeds.

4656. If the volume of air passing through a carburetor venturi is reduced, the pressure at the venturi throat will

1—decrease.
2—be equal to the pressure at the venturi inlet.
3—be equal to the pressure at the venturi outlet.
4—increase.

4657. Which curve shown in Figure 3, Fuel/Air Ratio Graphs, most nearly represents an aircraft engine's fuel/air ratio throughout its operating range?

1—A.
2—C.
3—D.
4—B.

4658. What will occur if the vapor vent float in a pressure carburetor loses its buoyancy?

1—A lean mixture will occur at all engine speeds.
2—The amount of fuel returning to the fuel tank from the carburetor will be increased.
3—The engine will continue to run after the mixture control is placed in IDLE CUTOFF.
4—A rich mixture will occur at all engine speeds.

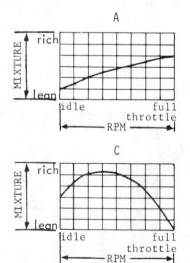

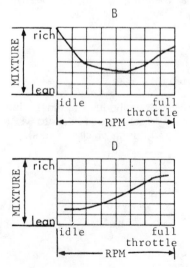

Figure 3

4659. What method is ordinarily used to make idle speed adjustments on a float–type carburetor?

1—An adjustable throttle stop or linkage.
2—An orifice and adjustable tapered needle.
3—An adjustable needle in the drilled passageway which connects the airspace of the float chamber and the carburetor venturi.
4—A variable restrictor in the idle system fuel supply.

4660. For what primary purpose is a turbine engine fuel control unit trimmed?

1—To obtain new exhaust gas temperature limits.
2—To obtain maximum thrust output when desired.
3—To properly position the power levers.
4—To adjust the idle RPM.

4661. Which type of fuel control is used on most of today's turbine engines?

1—Electromechanical.
2—Mechanical.
3—Hydromechanical.
4—Electronic.

4662. Under which of the following conditions will the trimming of a turbine engine be most accurate?

1—Low moisture and a tail wind.
2—High wind and high moisture.
3—High moisture and low wind.
4—No wind and low moisture.

4663. Due to the many variations in the engine's requirements, the fuel/air ratio varies with brake horsepower. The mixture used at rated power is

1—leaner than the mixture used at low horsepower and in idling ranges.
2—richer than the mixture used through the normal cruising range and in idling ranges.
3—richer than the mixture used at lower power settings above idle.
4—richer than the mixture used through the normal cruising range and leaner than the idle mixture.

4664. Under which of the following conditions would an engine run lean even though there is a normal amount of fuel present?

1—Engine operated at idle RPM.
2—The use of too high an octane rating fuel.
3—Incomplete fuel vaporization.
4—The carburetor air heater valve in the HOT position.

4665. During idle mixture adjustments, which of the following is normally observed to determine when the correct mixture has been achieved?

1—Changes in fuel/air pressure ratio.
2—Fuel flowmeter.
3—Fuel pressure indicator.
4—Changes in RPM or manifold pressure.

4666. The desired idle speed and mixture setting is one that will give

1—minimum RPM with maximum manifold pressure.
2—minimum RPM with minimum manifold pressure.
3—maximum RPM with minimum manifold pressure.
4—maximum RPM with maximum manifold pressure.

4667. The use of less than normal throttle opening during starting will cause

1—a rich mixture.
2—a lean mixture.
3—backfire due to lean fuel/air ratio.
4—preignition.

4668. When checking the idle mixture on a carburetor, the engine should be idling normally, then pull the mixture control toward the IDLE CUTOFF position. A correct idling mixture will be indicated by

1—an immediate decrease in RPM.
2—no change in RPM.
3—a decrease of 20 to 30 RPM before increasing.
4—an increase of 10 to 50 RPM before decreasing.

4669. When a new carburetor is installed on an engine,

1—warm up the engine and adjust the float level.
2—don't adjust the idle mixture setting; this was accomplished on the flow bench.
3—and the engine is warmed up to normal temperatures, adjust the idle mixture, then the idle speed.
4—the cruise mixture has to be adjusted after the first 10 hours of operation.

4670. The purpose of the back–suction mixture control in a float–type carburetor is to adjust the mixture by

1—regulating the pressure drop at the venturi.
2—regulating the pressure on the fuel in the float chamber.
3—regulating the suction on the mixture from behind the throttle valve.
4—restricting part of the fuel at all RPM settings.

4671. Engine power will be decreased at all altitudes if the

1—air density is increased.
2—humidity is increased.
3—manifold pressure is increased.
4—free–air temperature is decreased.

4672. If the idling jet becomes clogged in a float–type carburetor, the

1—engine operation will not be affected at any RPM.
2—engine will not idle.
3—accelerating pump will not operate.
4—idle mixture becomes richer.

4673. An aircraft engine, equipped with a pressure–type carburetor, is started with the

1—primer while the mixture control is positioned at IDLE CUTOFF.
2—mixture control positioned at the AUTO–RICH position.
3—mixture control in the FULL–RICH position.
4—primer while the mixture control is positioned at the FULL–LEAN position.

4674. One of the best ways to increase engine power and control detonation and preignition is to

1—enrich the fuel/air mixture.
2—use water injection.
3—lean the fuel/air mixture.
4—increase the carburetor throttle valve setting.

4675. The pressure drop at the venturi is very important in the operation of a float–type or pressure–type carburetor. The drop in pressure at the venturi depends primarily on the intake air

1—velocity.
2—weight.
3—humidity.
4—temperature.

4676. An excessively lean fuel/air mixture may cause

1—an increase in cylinder head temperature.
2—high oil pressure.
3—backfiring through the exhaust.
4—an increase in engine power.

4677. The density of air is very important when mixing fuel and air to obtain a correct fuel–to–air ratio. Which of the following weighs the most?

1—98 parts of dry air and two parts of water vapor.
2—75 parts of dry air and 25 parts of water vapor.
3—100 parts of dry air.
4—50 parts of dry air and 50 parts of water vapor.

4678. A fuel/air mixture ratio of 11:1 is

1—one part fuel to 11 parts air.
2—too rich to burn.
3—one part air to 11 parts fuel.
4—a lean mixture.

4679. The economizer system in a float–type carburetor

1—adds fuel to idle induction during sudden throttle movements.
2—keeps the fuel/air ratio constant.
3—functions only at cruise and idle speeds.
4—increases the fuel/air ratio at high power settings.

4680. A carburetor is prevented from leaning out during quick acceleration by the

1—power enrichment system.
2—boost venturi system.
3—mixture control system.
4—accelerating system.

4681. What are the positions of the pressurization valve and the dump valve in a jet engine fuel system during the engine starting cycle?

1—Pressurization valve open, dump valve open.
2—Pressurization valve closed, dump valve closed.
3—Pressurization valve closed, dump valve open.
4—Pressurization valve open, dump valve closed.

4682. What effect does high atmospheric humidity have on the operation of a jet engine?

1—Decreases P4 and increases P2.
2—Decreases compressor and turbine RPM.
3—Decreases fuel flow and increases EGT.
4—Has little or no effect.

4683. What are the positions of the pressurization valve and the dump valve in a jet engine fuel system when the engine is shut down?

1—Pressurization valve open, dump valve closed.
2—Pressurization valve closed, dump valve open.
3—Pressurization valve open, dump valve open.
4—Pressurization valve closed, dump valve closed.

4684. Which of the following may cause a lean mixture and high cylinder temperature?

1—Ruptured balance diaphragm.
2—Defective automatic mixture control unit.
3—Leaking primer.
4—Defective accelerating system.

4685. Three parameters monitored in a jet engine by the fuel control unit are

1—compressor inlet pressure, nozzle pressure, and burner pressure.
2—compressor outlet pressure, tail pipe temperature, and compressor RPM.
3—compressor inlet pressure, compressor RPM, and turbine inlet pressure.
4—compressor inlet pressure, burner pressure, and compressor RPM.

4686. Which component, within a fuel control unit, prevents compressor stall during rapid acceleration?

1—Limiting valve.
2—Inlet pressure sensor.
3—Turbine pressure sensor.
4—Override check valve.

4687. Detonation causes high cylinder head temperature because the fuel mixture

1—burns too fast.
2—burns too slow.
3—is too rich.
4—is too lean.

4688. What corrective action should be taken when a carburetor is found to be leaking fuel from the discharge nozzle?

1—Replace the needle valve and seat.
2—Raise the float level.
3—Turn the fuel off each time the aircraft is parked.
4—Replace the gasket between the float bowl and the throttle body.

4689. During what period does the fuel pump bypass valve open and remain open?

1—When the wobble pump is operated in a parallel–type fuel system.
2—When the fuel pump pressure is greater than the demand of the engine.
3—When the boost pump pressure is greater than fuel pump pressure.
4—When the fuel pump output is greater than the demand of the carburetor.

4690. One of the following statements concerning a centrifugal–type fuel boost pump located in a fuel supply tank is not true.

1—Air and fuel vapors do not pass through a centrifugal–type pump.
2—Fuel can be drawn through the impeller section of the pump when it is not in operation.
3—The discharge side of the pump supplies fuel to the engine–driven pump.
4—The centrifugal–type pump is classified as a positive displacement pump.

4691. Where is the engine fuel shutoff valve usually located?

1—Aft of the firewall.
2—Adjacent to the carburetor.
3—Adjacent to the fuel pump.
4—Downstream of the engine–driven fuel pump.

4692. Boost pumps in a fuel system

1—operate during takeoff only.
2—are primarily used for fuel transfer.
3—are only in secondary tanks.
4—provide a positive flow of fuel to the engine pump.

4693. What is the purpose of the fuel transfer ejectors as shown in Figure 4, Fuel System?

1—To supply fuel under pressure to the engine–driven pump.
2—To assist in the transfer of fuel from the main tank to the boost pump sump.
3—To transfer fuel from the boost pump sump to the wing tank.
4—To assist in the transfer of fuel during climb.

4694. What is the purpose of an engine–driven fuel pump bypass valve?

1—To divert the excess fuel back to the main tank.
2—To prevent a damaged or inoperative pump from blocking the fuel flow of another pump in series with it.
3—To prevent excessive fuel pressure at the fuel inlet of the carburetor.
4—To divert the excess fuel from the pressure side of the pump to the inlet side of the pump.

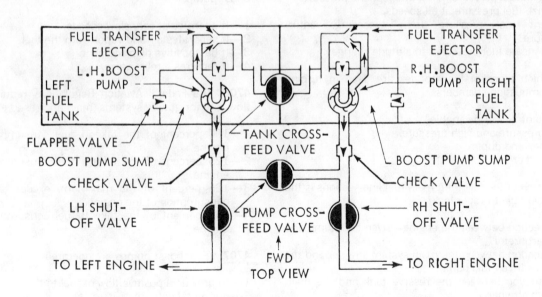

Figure 4

4695. Most large aircraft reciprocating engines are equipped with which of the following types of engine-driven fuel pumps?

1—Rotary-vane-type fuel pump.
2—Wobble-type fuel pump.
3—Centrifugal-type fuel pump.
4—Gear-type fuel pump.

4696. When an electric primer is used, fuel pressure is built up by the

1—internal pump in the primer solenoid.
2—carburetor accelerating pump.
3—suction at the main discharge nozzle.
4—booster pump.

4697. The fuel pump relief valve directs excess fuel to the

1—fuel tank return line.
2—inlet side of the fuel pump.
3—inlet side of the fuel strainer.
4—fuel pump drain line.

4698. Which type of pump is commonly used as a fuel pump on reciprocating engines?

1—Gear.
2—Wobble.
3—Impeller.
4—Vane.

4699. The purpose of the diaphragm in most vane-type fuel pumps is to

1—maintain fuel pressure below atmospheric pressure.
2—equalize fuel pressure at all speeds.
3—prevent fuel pressure from exceeding venturi pressure.
4—compensate fuel pressures to altitude changes.

4700. Which types of fuel nozzles are used in turbine engine combustion chambers?

1—Constant flow and variable.
2—Low pressure and high pressure.
3—Simplex and duplex.
4—Idle and cruise.

4701. Which of the following main fuel line sections is the least susceptible to vapor lock?

1—The section between the engine-driven fuel pump and the carburetor.
2—The section between the main system strainer and the engine-driven fuel pump.
3—The section between the reserve tank and the main system strainer
4—The section between the main tank and the reserve tank.

4702. The fuel systems of aircraft certificated in the standard classification must include which of the following?

1—An engine-driven fuel pump and at least one auxiliary pump per engine.
2—An acceptable method for indicating the rate of fuel consumption for each engine.
3—A positive means of shutting off the fuel to all engines.
4—A reserve supply of fuel, available to the engine only after selection by the flightcrew, sufficient to operate the engines at least 30 minutes at METO power.

4703. Where should the main fuel strainer be located in the aircraft fuel system?

1—Downstream from the wobble pump check valve.
2—Downstream from the carburetor strainer.
3—At the lowest point in the fuel system.
4—At a point in the system lower than the carburetor strainer, but higher than the finger strainer.

4704. Where physical separation of the fuel lines from electrical wiring or conduit is impracticable, locate the fuel line

1—below the wiring and clamp the line securely to the airframe structure.
2—adjacent to the wiring and clamp them together.
3—above the wiring and clamp the line securely to the airframe structure.
4—inboard of the wiring and clamp both securely to the airframe structure.

4705. What is a characteristic of a centrifugal-type fuel boost pump?

1—It operates at very slow speeds.
2—It separates air and vapor from the fuel.
3—It has positive displacement.
4—It requires a relief valve.

4706. The Federal Aviation Regulations require the fuel flow rate for gravity systems (main and reserve) to be

1—100 percent of the takeoff fuel consumption of the engine.
2—125 percent of the takeoff fuel consumption of the engine.
3—125 percent of the maximum, except takeoff, fuel consumption of the engine.
4—150 percent of the takeoff fuel consumption of the engine.

4707. Fuel boost pumps are operated

1—to provide a positive flow of fuel to the engine.
2—during takeoff only.
3—primarily for fuel transfer to another tank.
4—automatically from fuel to the pressure source switch.

4708. A pilot reports that the fuel pressure fluctuates and exceeds the upper limits whenever the throttle is advanced. The most likely cause of the trouble is

1—a ruptured fuel pump relief–valve diaphragm.
2—a sticky fuel pump relief valve.
3—loose bolts in the fuel pump body to the relief–valve housing.
4—an air leak at the fuel pump relief–valve body.

4709. A fuel strainer or filter must be located between the

1—boost pump and tank outlet.
2—carburetor fuel chamber and throttle body.
3—tank outlet and the fuel metering device.
4—engine–driven fuel pump and vapor vent line.

4710. Fuel pump relief valves designed to compensate for atmospheric pressure variations are known as

1—thermo–pressure relief valves.
2—compensated–flow valves.
3—pressurized–relief valves.
4—balanced–type relief valves.

4711. Fuel lines are kept away from sources of heat and sharp bends and steep rises are avoided to reduce the possibility of

1—liquid lock.
2—vapor lock.
3—air lock.
4—positive lock.

4712. Fuel crossfeed systems are used in aircraft to

1—defuel tanks.
2—purge the fuel tanks.
3—jettison fuel in case of an emergency.
4—maintain aircraft stability.

4713. A float–type carburetor used with an engine–driven pump system will most likely have a fuel pressure of about

1—1/2 pound.
2—6 to 7 pounds.
3—12 to 15 pounds.
4—3–1/2 to 5 pounds.

4714. A fuel pressure relief valve is required on

1—a wobble pump.
2—a centrifugal fuel boost pump.
3—an engine–driven fuel pump.
4—a main fuel strainer.

4715. A rotary–vane pump is best described as

1—a positive–displacement pump.
2—a variable–displacement pump.
3—a boost pump.
4—an auxiliary pump.

4716. Fuel pressure produced by the engine–driven fuel pump is adjusted by the

1—bypass valve adjusting screw.
2—relief valve adjusting screw.
3—main fuel strainer adjusting screw.
4—engine–driven fuel pump adjusting screw.

4717. A vane–type fuel pump is used

1—on most aircraft as a booster pump.
2—only in fuel systems for helicopters.
3—as the engine–driven fuel pump on most aircraft.
4—in most fuel systems because a relief valve is not required, since it is not a positive displacement pump.

4718. Gasoline and kerosene are used as turbine engine fuels. How do they compare in heat energy?

1—Gasoline has more heat energy per gallon than kerosene.
2—Kerosene has more heat energy per gallon than gasoline.
3—Gasoline and kerosene have the same heat energy per unit of volume.
4—Gasoline and kerosene have the same heat energy per unit of weight.

4719. What are the principle advantages of the duplex fuel nozzle used in most turbine engines?

1—Restricts the amount of fuel flow to a level where more efficient and complete burning of the fuel is achieved.
2—Provides better atomization and uniform flow pattern.
3—Allows a wider range of fuels and filters to be used.
4—Requires less filtering and blending of the fuel because of the dual passages in the nozzle head.

4720. Why is it necessary to control acceleration and deceleration rates of turbine engines?

1—To prevent blow–out.
2—To prevent overtemping.
3—To enable the engine to heat and cool at controlled rates.
4—To prevent friction between turbine wheels and the case due to expansion and contraction.

4721. When trimming a turbine engine, the fuel control is adjusted to

1—produce as much power as the engine is capable of producing.
2—limit idle RPM and maximum speed.
3—allow the engine to produce maximum RPM without regard to power output.
4—restrict power to 100 percent without regard to specified RPM.

4722. Which of the following turbine fuel filters has the greatest filtering action?

1—Micron.
2—Wire mesh.
3—Wafer screen.
4—Stocked charcoal.

4723. What is the purpose of the flow divider in a turbine engine duplex fuel nozzle?

1—Allows an alternate flow of fuel if the primary flow clogs or is restricted.
2—Directs excessive fuel back to the fuel manifold.
3—Creates the primary and secondary fuel supplies.
4—Provides a flow path for bleed air which aids in the atomization of fuel.

4724. What causes the fuel divider valve to open in a turbine engine duplex fuel nozzle?

1—Fuel pressure.
2—Thermostatically controlled heat from the combustion section.
3—Bleed air after the engine reaches idle RPM.
4—An electrically operated solenoid.

4725. A method commonly used to prevent carburetor icing is to

1—preheat the intake air.
2—mix alcohol with the fuel.
3—coat the butterfly valve with glycerine.
4—use water.

4726. The most generally accepted method for controlling carburetor icing is to

1—preheat the fuel to the carburetor.
2—add a pint of alcohol to every 5 gallons of fuel.
3—maintain the temperature of the fuel/air mixture just above freezing point of water.
4—warm the carburetor intake air.

4727. Carburetor icing is most severe at

1—air temperatures between 30° F and 40° F.
2—high altitudes.
3—low engine temperatures.
4—air temperatures below 0° F.

4728. Into what part of a reciprocating engine induction system is deicing alcohol normally injected?

1—The supercharger or impeller section.
2—The airstream ahead of the carburetor.
3—The low-pressure area ahead of the throttle valve.
4—The fuel ahead of the discharge nozzle.

4729. Carburetor icing on an engine equipped with a constant-speed propeller can be detected by

1—a decrease in power output with no change in manifold pressure or RPM.
2—an increase in manifold pressure with a constant RPM.
3—a decrease in manifold pressure with a constant RPM.
4—a decrease in manifold pressure with a decrease in RPM.

4730. What part of an aircraft in flight will begin to accumulate ice before any other?

1—Wing leading edge.
2—Propeller spinner or dome.
3—Nose or fuselage on multiengine aircraft.
4—Carburetor.

4731. Carburetor icing may be eliminated by which of the following methods?

1—Alcohol spray and electrically heated induction duct.
2—Ethylene glycol spray and heated induction air.
3—Alcohol spray and heated induction air.
4—Electrically heated air intake and ethylene glycol spray.

4732. Where would a carburetor air heater be located in a fuel injection system?

1—Between the air intake and the cylinders.
2—At the air intake entrance.
3—None is required.
4—Between the air intake and the venturi.

4733. An increase in manifold pressure when carburetor heat is applied indicates

1—excessive heat is being used.
2—ice was forming in the carburetor.
3—mixture was too lean.
4—overheating of cylinder heads.

4734. During full power output of an unsupercharged engine equipped with a float-type carburetor, in which of the following areas will the highest pressure exist?

1—Engine side of throttle valve.
2—Venturi.
3—Intake manifold.
4—Carburetor air scoop.

4735. The use of the carburetor air heater when it is not needed would cause

1—very lean mixture.
2—excessive increase in manifold pressure.
3—a decrease in power and possibly detonation.
4—damage to the carburetor from excessive heat.

4736. Which of the following will indicate normal operation of a single-stage, two-speed supercharger, when shifted from low to high impeller ratio during a ground check?

1—An increase in RPM, a momentary drop in oil pressure, and an increase in manifold pressure.
2—Oil and manifold pressure remain unchanged with an increase in RPM.
3—A drop in oil and manifold pressure with a decrease in RPM.
4—A drop in manifold pressure, a drop in oil pressure, RPM unchanged with an increase in fuel pressure.

4737. Which of the following is a function of the diffuser of an internal supercharger?

1—It assures uniform distribution of fuel/air mixture to the cylinders.
2—It decreases pressure of fuel/air mixture.
3—It increases the velocity of inlet air.
4—It introduces more air into the mixture.

4738. As manifold pressure increases in a reciprocating engine, the

1—volume of air in the cylinder increases.
2—weight of the fuel/air charge decreases.
3—density of air in the cylinder increases.
4—volume of air in the cylinders decreases.

4739. Which of the following statements regarding volumetric efficiency of an engine is true?

1—The volumetric efficiency of an engine will remain the same regardless of the amount of throttle opening.
2—It is impossible to exceed 100 percent volumetric efficiency of any engine regardless of the type of supercharger used.
3—Manifold pressure will increase as altitude horsepower is increased.
4—It is possible to exceed 100 percent volumetric efficiency of some engines by the use of internal superchargers of the proper type.

4740. If an engine is equipped with an external turbocharger and is started with the waste gate closed, the result might be

1—some damage to cores of the turbocharger intercooler.
2—overspeeding of the turbocharger with resultant damage to pistons and rings.
3—serious damage because of overboost.
4—nothing; the engine should be started with the waste gate closed.

4741. Diffuser vanes located in the supercharger section of a radial engine are designed to perform one of the following functions.

1—Increase the pressure of the fuel/air charge.
2—Increase the velocity of the fuel/air charge.
3—Decrease the pressure of the fuel/air charge.
4—Increase the temperature of the fuel/air charge.

4742. Which of the following would be a factor in the failure of an engine to develop full power at takeoff?

1—Failure to install the carburetor scoop air screen.
2—Improper adjustment of carburetor heat valve control linkage.
3—Excessively rich setting on the idle mixture adjustment.
4—Failure of the economizer valve to remain closed at takeoff throttle setting.

4743. If the turbosupercharger waste gate is completely closed,

1—none of the exhaust gases are directed through the turbine.
2—the manifold pressure will be lower than normal.
3—the turbosupercharger is in the OFF position.
4—all the exhaust gases are directed through the turbine.

4744. Boost manifold pressure is generally considered to be any manifold pressure above

1—14.7 inches Hg.
2—50 inches Hg.
3—40 inches Hg.
4—30 inches Hg.

4745. What is the purpose of the density controller in a turbocharger system?

1—Maintains constant air velocity at the carburetor venturi.
2—Limits the maximum manifold pressure that can be produced at other than full throttle conditions.
3—Limits the maximum manifold pressure that can be produced by the turbocharger at full throttle.
4—Maintains constant air velocity at the carburetor inlet.

4746. What is the purpose of the rate-of-change controller in a turbocharger system?

1—Limits the maximum manifold pressure that can be produced by the turbocharger at full throttle conditions.
2—Limits the maximum manifold pressure that can be produced at other than full throttle conditions.
3—Controls the rate at which the turbocharger discharge pressure will increase.
4—Controls the position of the waste gate after the aircraft has reached its critical altitude.

4747. What regulates the speed of a turbosupercharger?

1—Turbine.
2—Compressor.
3—Waste gate.
4—Throttle.

4748. What is the purpose of a turbocharger system for a small reciprocating aircraft engine?

1—Compresses the air to hold the cabin pressure constant after the aircraft has reached its critical altitude.
2—Maintains constant air velocity in the intake manifold.
3—Compresses air to maintain manifold pressure constant from sea level to the critical altitude of the engine.
4—Maintains variable air pressure to the carburetor venturi.

4749. What are the three basic regulating components of a sea–level boosted turbocharger system?

A. The exhaust bypass assembly.
B. The compressor assembly.
C. The pump and bearing casing.
D. The density controller.
E. The differential pressure controller.

1—B, C, D.
2—A, D, E.
3—C, D, E.
4—A, B, C.

4750. The differential pressure controller in a turbocharger system

1—reduces bootstrapping during part–throttle operation.
2—positions the waste gate valve for maximum power.
3—provides a constant fuel to air ratio.
4—positions the waste gate valve to minimize exhaust back pressure.

4751. What is the purpose of a pressure ratio controller in a turbocharger system?

1—Controls the position of the waste gate after the aircraft has reached its critical altitude.
2—Controls the position of the waste gate between sea level and the critical altitude of the aircraft.
3—Controls the rate at which the turbocharger discharge pressure is allowed to change.
4—Controls the air which holds the manifold pressure constant from sea level to the critical altitude of the engine.

4752. What is used to drive an externally driven supercharger?

1—Engine oil pressure.
2—Gear driven directly from the engine crankshaft.
3—Exhaust gases driving a turbine.
4—Belt driven through a pully arrangement.

4753. What are the three parts of a typical turbosupercharger for a large reciprocating engine?

A. Density controller.
B. Compressor assembly.
C. Differential pressure controller.
D. Exhaust gas turbine assembly.
E. Pump and bearing casing.

1—A, B, C.
2—B, D, E.
3—A, C, E.
4—C, D, E.

4754. If carburetor or induction system icing is not present when carburetor heat is applied with no change in the throttle setting, the

1—mixture will become richer.
2—manifold pressure will increase.
3—engine power output will increase.
4—engine RPM will increase.

4755. When starting an engine equipped with a carburetor air heater, in what position should the heater be placed?

1—Hot.
2—Cold.
3—Halfway open.
4—Neutral.

4756. The application of carburetor heat during engine operation will

1—decrease the weight of the fuel/air charge.
2—decrease the volume of air in the cylinder.
3—increase the volume of air in the cylinder.
4—increase the density of air in the cylinder.

4757. The application of carburetor heat will have which of the following effects?

1—The manifold pressure will be increased.
2—The mixture will become leaner.
3—Less throttle opening will be required for the same power.
4—The mixture will become richer.

4758. When operating an engine, the application of carburetor heat will have what effect on the fuel/air mixture?

1—Enriching the mixture because the AMC cannot make a correction for increased temperature.
2—Leaning the mixture because the AMC cannot make a correction for increased temperature.
3—Enriching the mixture until the AMC can make a compensation.
4—Leaning the mixture until the AMC can make a compensation.

4759. Electrical priming systems usually obtain fuel from the

1—carburetor.
2—oil dilution solenoid.
3—blower throat.
4—engine–driven pump.

4760. On a multiple–point priming system used on a nine–cylinder radial engine, priming lines will be connected to

1—cylinders Nos. 4, 5, 6, and 7.
2—cylinders Nos. 1, 3, 5, 7, and 9.
3—cylinders Nos. 1, 2, 3, 8, and 9.
4—cylinders Nos. 1, 2, 3, 4, 5, 6, 7, 8, and 9.

4761. Induction fires during aircraft starting that continue to burn may be extinguished by

1—shutting the engine off and putting out the fire with available equipment.
2—continuing to run engine to extinguish the fire.
3—calling the Flight Service Station for service in extinguishing the fire.
4—alerting the tower and requesting their aid in extinguishing the fire.

4762. If a fire starts in the induction system during the engine starting procedure, what should the operator do?

1—Turn off the fuel switches to stop the fuel.
2—Continue cranking the engine.
3—Backfire the engines to blow the fire out.
4—Turn off all switches.

4763. On small aircraft engines, fuel vaporization may be increased by

1—cooling the air before it enters the engine.
2—circulating the fuel and air mixture through passages in the oil sump.
3—heating the fuel before it enters the carburetor.
4—routing the exhaust gas around the fuel lines.

4764. The action of a carburetor airscoop is to supply air to the carburetor, but it may also

1—prevent ice formation.
2—cool the engine.
3—keep fuel lines cool and prevent vapor lock.
4—increase the pressure of the incoming air by ram effect.

4765. A carburetor pre–heater is not generally used on takeoff unless absolutely necessary because of the

1—loss of power and possible detonation.
2—drain on the aircraft electrical system.
3—fire hazard involved.
4—inability of the engine to supply enough heat to make any difference.

4766. Baffles and deflectors are installed around cylinders of air–cooled aircraft engines to

1—force cooling air into close contact with all parts of the cylinder.
2—produce a low–pressure area that will decrease the air velocity of the cooling air.
3—create a high pressure in front of the cylinders.
4—prevent damage to the cylinder head and barrel cooling fins.

4767. What is the purpose of an augmentor used in some reciprocating engine exhaust systems?

1—To reduce exhaust back pressure.
2—To aid in cooling the engine.
3—To assist in displacing the exhaust gases.
4—To augment the surface area of the exhaust extension.

4768. Aircraft reciprocating engine cylinder baffles and deflectors should be repaired as required to prevent loss of

1—power.
2—fin area.
3—carburetor air.
4—cooling.

4769. Cylinder baffles and deflectors

1—provide for more uniform cooling of the cylinders.
2—have no special purpose.
3—prevent damage to fins.
4—improve the looks of the engine.

4770. If cracks in cooling fins do not extend into the cylinder head, they may be repaired by

1—stop drilling extremities of cracks in the head portion.
2—removing affected area and contour filing within limits.
3—welding or brazing, preferably brazing.
4—complete coverage by appropriate metalizing within limits.

4771. Which of the following should a mechanic consult to determine the maximum amount of cylinder cooling fin that could be removed when re–profiling?

1—Advisory Circular 43.13–1A.
2—Engine manufacturer's service or overhaul manual.
3—Federal Aviation Regulation, Part 43, Appendix A.
4—Advisory Circular 43.13–2A.

4772. A bent cooling fin on an aluminum cylinder head

1—will cause rejection of the cylinder.
2—should be sawed off and filed smooth.
3—should be left alone if no crack has formed.
4—should be stop drilled or a small radius filed at the point of the bend.

4773. Where are cooling fins usually located on air–cooled engines?

1—Exhaust side of cylinder head, connecting rods, and cylinder walls.
2—Exhaust side of the cylinder head, inside the pistons, and connecting rods.
3—Cylinder head, cylinder walls, and inside the piston skirt.
4—Cylinder head, cylinder barrel, and inside the piston head.

4774. Which of the following assists in removing heat from the metal walls and fins of an air–cooled cylinder assembly?

1—An integral heat pump.
2—An intercooler system.
3—A baffle and cowl arrangement.
4—A heat muff exchanger.

4775. During ground operation of an engine, the cowl flaps should be in what position?

1—Fully closed.
2—Fully open.
3—One–third open.
4—Two–thirds open.

4776. Turbine engines that depend solely on lubricating oil for bearing cooling normally use

1—an oil cooler in the system.
2—a larger capacity oil pressure pump.
3—a larger capacity oil tank.
4—a small quantity of oil.

4777. During an operational check of an electrically powered radial engine cowl flap system, the motor fails to operate. Which of the following is the first to be checked?

1—Flap actuator motor circuit breaker.
2—The cockpit control switch.
3—Flap actuator jackscrew synchronization switch.
4—Flap actuator motor.

4778. (1) Some aircraft exhaust systems include an augmentor system to draw additional cooling air over the engine.

(2) Augmentor system may provide heated air for cabin heat, anti–icing and deicing systems.

1—Neither No. 1 nor No. 2 is true.
2—Only No. 1 is true.
3—Both No. 1 and No. 2 are true.
4—Only No. 2 is true.

4779. Which of the following defects would likely cause a hot spot on a reciprocating engine cylinder?

1—Too much cooling fin area broken off.
2—A cracked cylinder baffle.
3—A cracked cylinder baffle blast tube.
4—Cowling air seal leakage.

4780. What part of an air–cooled cylinder assembly has the greatest fin area per square inch?

1—The cylinder barrel.
2—The rear of the cylinder head.
3—The junction of the cylinder barrel and head.
4—The exhaust valve port.

4781. Reciprocating engines used in helicopters are cooled by

1—cowl flaps.
2—the downdraft from the main rotor.
3—a fan mounted on the engine.
4—blast tubes on either side of the engine mount.

4782. The greatest portion of heat generated by combustion in a typical aircraft reciprocating engine is

1—converted into useful power.
2—removed by the oil system.
3—carried out with the exhaust gases.
4—dissipated through the cylinder walls and heads.

4783. The purpose of baffle plates around air–cooled cylinders is to

1—provide for more uniform and efficient cooling.
2—force air through the cabin heater system.
3—protect the fins from damage.
4—obtain a better streamlined effect.

4784. Cracks at the base of a cooling fin on a cylinder

1—may be stop drilled.
2—are cause for rejection of cylinder.
3—may be acceptable.
4—may be filed and profiled according to AC 43.13–1A.

4785. An engine becomes overheated due to excessive taxiing or improper ground runup. Prior to shutdown, operation must continue until cylinders have cooled, by running engine at

1—low RPM with oil dilution system activated.
2—idle RPM.
3—high RPM with mixture control in lean position.
4—high RPM with mixture control in rich position.

4786. Cylinder head temperatures are measured by means of an indicator and a

1—resistance bulb sensing device.
2—wheatstone bridge sensing device.
3—thermocouple sensing device.
4—Bourdon tube sensing device.

4787. The rows of cylinders on a twin–row radial engine are staggered with respect to each other

1—to facilitate cooling.
2—to permit the use of one main dynamic damper.
3—so that one cam gear can operate both rows.
4—so one magneto can fire the front plugs on both rows.

4788. High cylinder head temperature will result from

1—an increase in oil consumption.
2—fouled spark plugs.
3—an excessively rich mixture.
4—an excessively lean mixture.

4789. The purpose of an intercooler when used with a turbosupercharger is to

1—cool the exhaust gases before they come in contact with the turbo drive.
2—cool the supercharger bearings.
3—cool the mixture of fuel and air entering the internal supercharger.
4—cool the air entering the carburetor from the turbosupercharger.

4790. Prolonged idling of an engine will usually result in

1—excessive cylinder head temperatures.
2—burned magneto points.
3—excessive oil consumption.
4—foreign material buildup on spark plugs.

4791. Which of the following best describes a cylinder muff?

1—A device to absorb sound and reduce engine operation noise.
2—A heating shroud used to provide cabin heat.
3—A separate sleeve of aluminum cooling fins shrunk on to the steel inner cylinder sleeve.
4—A cooling jacket for liquid–cooled engines.

4792. What is the function of a blast tube as found on aircraft engines?

1—A means of cooling the engine by utilizing the propeller backwash.
2—A device to indicate airspeed.
3—A tube used to load a cartridge starter.
4—A device to cool an engine accessory.

4793. The air passing through the combustion chamber of a jet engine is

1—used to support combustion and to cool the engine.
2—speeded up and heated by the action of the turbines.
3—entirely combined with the fuel and burned.
4—heated by the fuel and expanded, but otherwise unchanged.

4794. Which of the following results in a decrease in volumetric efficiency?

1—Cylinder head temperature too low.
2—Carburetor air temperature too low.
3—Part–throttle operation.
4—Short intake pipes of large diameter.

4795. The undersides of pistons are frequently finned. The principal reason is to

1—provide sludge chambers and sediment traps.
2—provide for greater heat transfer to the engine oil.
3—support ring grooves and piston pins.
4—support the piston pin bosses.

4796. What is the position of the cowl flaps during engine starting and warmup operations under normal conditions?

1—Full open at all times.
2—Full closed at all times.
3—Open for starting, closed for warmup.
4—An intermediate position.

4797. Increased engine heat will cause volumetric efficiency to

1—remain the same.
2—decrease or increase depending on the engine RPM.
3—decrease.
4—increase.

4798. Why is high nickel chromium steel used in many exhaust systems?

1—Low expansion coefficient and high flexibility.
2—High heat conductivity and flexibility.
3—Corrosion resistance and low expansion coefficient.
4—Corrosion resistance and high heat conductivity.

4799. Slip joints are required in most exhaust collector systems because of the

1—high heat conductivity of metal causing distortion.
2—installation requirements.
3—necessity of installing the unit piece by piece.
4—difficulty in aligning mounting bolts.

4800. A carburetor air intake heater which removes ice from the induction system usually derives its heat from the

1—electric heating elements.
2—cabin heater.
3—gasoline or alcohol flame.
4—exhaust gases.

4801. The hot section of a turbine engine is particularly susceptible to which of the following kind of damage?

1—Galling.
2—Pitting.
3—Cracking.
4—Scoring.

4802. What is the purpose of a slip joint in an exhaust collector ring?

1—It aids in alignment and absorbs expansion.
2—It reduces vibration and increases cooling.
3—It permits the collector ring to be installed in one piece.
4—It increases service life of collector ring segments.

4803. Sodium filled valves are advantageous to an aviation engine because they

1—cost less, as they run hotter.
2—are lighter.
3—have great strength properties.
4—dissipate heat well.

4804. What type nuts are used to hold an exhaust system to the cylinders?

1—Brass or special locknuts.
2—High–temperature fiber self–locking nuts.
3—Low–temperature steel self–locking nuts.
4—High–temperature aluminum self–locking nuts.

4805. What are two general types of exhaust systems in use on reciprocating engines?

A. Bifurcated duct arrangement.
B. Short stacks.
C. Long stacks.
D. Collector system.
E. Shroud system.

1—A and C.
2—B and D.
3—D and E.
4—A and E.

4806. Repair of exhaust system components

1—is impossible because the material cannot be identified.
2—must be accomplished by the component manufacturer.
3—is usually accomplished using fiberglass patch kits.
4—is not recommended to be accomplished in the field.

4807. On turbojet–powered airplanes, how much reverse thrust is usually required for minimum braking requirements?

1—At least 50 percent of the full forward thrust of the engine.
2—At least 100 percent of the full forward thrust of the engine.
3—At least 75 percent of the full forward thrust of the engine.
4—At least 25 percent of the full forward thrust of the engine.

4808. On an aircraft that utilizes an exhaust heat exchanger as a source of cabin heat. how should the exhaust system be inspected?

1—X–rayed to detect any cracks.
2—Tested by use of an exhaust gas analyzer.
3—Hydrostatically tested.
4—With the heater air shroud removed.

4809. How should ceramic–coated exhaust components be cleaned?

1—By sandblasting.
2—With alkali.
3—By degreasing.
4—By mechanical means.

4810. Which of the following indicates that a combustion chamber of a jet engine is not operating properly?

1—Clam shells stick in thrust reverse position.
2—Hot spots on the tail cone.
3—Missing teeth in the synchronizing gear segment.
4—Warping of the clam shells.

4811. Select a characteristic of a good gas weld on exhaust stacks.

1—The weld should be built up 1/8 inch.
2—Heavy oxide is formed on the base metal close to the weld.
3—Porousness or projecting globules should show in the weld.
4—The weld should taper off smoothly into the base metal.

4812. How do the turbines which are driven by the exhaust gases of a turbo–compound engine contribute to total engine power output?

1—By driving the crankshaft through suitable couplings.
2—By causing the exhaust back pressure to remain below atmospheric pressure at all operating altitudes.
3—By driving the supercharger, thus relieving the engine of the supercharging load.
4—By converting the latent heat energy of the exhaust gases into thrust by collecting and accelerating them.

4813. How should corrosion–resistant steel parts such as exhaust collectors be blast cleaned?

1—Use steel grit which has not previously been used on soft iron.
2—Use super fine granite grit.
3—Use sand which has not previously been used on iron or steel.
4—Use soft iron chill which has not previously been used on hardened steel.

4814. Power recovery turbines used on some reciprocating engines are driven by the

1—exhaust gas pressure.
2—crankshaft.
3—velocity of the exhaust gases.
4—fluid drive coupling.

4815. Reciprocating engine exhaust systems that have repairs or sloppy weld beads which protrude internally are unacceptable because they cause

1—base metal fatigue.
2—localized cracks.
3—unrestricted exhaust gas flow.
4—local hot spots.

4816. Ball joints in reciprocating engine exhaust systems should be

1—tight enough to prevent any movement.
2—disassembled and the seals replaced every engine change.
3—secured to each exhaust extension with AN bolts, plain nuts, and lockwashers.
4—loose enough to permit some movement.

4817. All of the following are recommended markers for reciprocating engine exhaust systems except

1—India ink.
2—chalk.
3—lead pencil.
4—Prussian blue.

4818. What is the function of the thrust reverser of a turbojet engine?

1—To extend the reverser flaps.
2—To reverse the flow of exhaust gases.
3—To reverse the airflow through engine inlet.
4—To reduce the velocity of exhaust gases.

4819. How are combustion liner walls cooled in a gas–turbine engine?

1—By secondary air flowing through the combustion chamber.
2—By the pattern of holes and louvers cut in the diffuser section.
3—By ram air from engine air intake.
4—By bleed air vented from the engine air inlet.

4820. Augmentor tubes are part of which reciprocating engine system?

1—Induction.
2—Oil.
3—Exhaust.
4—Fuel.

4821. Dislodged internal muffler baffles on a small reciprocating engine may

1—obstruct the muffler outlet and cause excessive exhaust back pressure.
2—cause the engine to run excessively cool.
3—cause high fuel and oil consumption.
4—result in an engine fire.

4822. What is the purpose of an exhaust outlet guard on a small reciprocating engine?

1—To prevent dislodged muffler baffles from obstructing the muffler outlet.
2—To reduce spark exit.
3—To protect the muffler outlet during ground servicing activities.
4—To shield adjacent components from excessive heat.

4823. What could be a result of undetected exhaust system leaks in a reciprocating engine powered airplane?

1—Pilot incapacitation resulting from carbon monoxide entering the cabin.
2—A rough–running engine.
3—Desired power settings will not be attained.
4—Excessive engine operating temperatures.

4824. How may reciprocating engine exhaust system leaks be detected?

1—An exhaust trail aft of the tailpipe on the airplane exterior.
2—Low cylinder head temperature indication.
3—Fluctuating manifold pressure indication.
4—Signs of exhaust soot inside cowling and on adjacent components.

4825. Reciprocating engine exhaust systems often operate at temperatures

1—in excess of 1,500° F.
2—at 2,000° F.
3—between 250° and 500° F.
4—near 1,000° F.

4826. Most exhaust system failures result from thermal fatigue cracking in the areas of stress concentration. This condition is usually caused by

1—the drastic temperature change which is encountered at altitude.
2—improper welding techniques during manufacture.
3—the low temperatures which the exhaust system is subjected to during initial warmup.
4—the high temperatures at which the exhaust system operates.

4827. How is aircraft electrical power for propeller deicer system transferred from the engine to the propeller hub assembly?

1—By slip rings and segment plates.
2—By slip rings and brushes.
3—By collector ring and transducer.
4—By flexible electrical connectors.

4828. How is anti–icing fluid ejected from the slinger ring on a propeller?

1—By ejector valves.
2—By pump pressure.
3—By centripetal force.
4—By centrifugal force.

4829. On most reciprocating multiengine aircraft, automatic propeller synchronization is accomplished through the actuation of the

1—blade switches.
2—throttle levers.
3—propeller governors.
4—propeller control levers.

4830. Propeller fluid anti–icing systems generally use which of the following?

1—Ethylene glycol.
2—Isopropyl alcohol.
3—Denatured alcohol.
4—Ethyl alcohol.

4831. Differences in the speed of two engines under sychronizer control are detected by

1—a master override relay.
2—a propeller governor.
3—blade switches.
4—a master motor.

4832. Which of the following is the most common fluid used for propeller anti–icing on a reciprocating engine aircraft?

1—Ethylene glycol.
2—Isopropyl alcohol.
3—Denatured alcohol.
4—Phosphate compounds.

4833. What is a function of the automatic propeller synchronizing system on multiengine aircraft?

1—To increase vibration and reduce noise.
2—To control the tip speed of all propellers.
3—To control engine RPM and reduce vibration.
4—To control the power output of all engines.

4834. Ice formation on propellers, when the aircraft is in flight, will

1—decrease thrust and cause excessive vibration.
2—increase stall speed and increase noise.
3—decrease stall speed and increase noise.
4—increase thrust and cause excessive vibration.

4835. Proper operation of a propeller deicing system can be monitored by observing which of the following?

1—Voltmeters.
2—Indicator lights.
3—Audible cycling device.
4—Ammeters.

4836. What unit in the propeller anti–icing system controls the output of the pump?

1—Pressure relief valve.
2—Rheostat.
3—Cycling timer.
4—Current limiter.

4837. Propeller electric deicing boots can be checked for operation by

1—observing the ammeter or loadmeter for current flow.
2—checking the ammeter for flickering and feeling the boots for sequence of heating.
3—timing the inflation and deflation sequence.
4—feeling them to determine if they are hot.

4838. Which of the following determines oil and grease specifications for lubrication of propellers?

1—FAA engineering.
2—Petroleum institute.
3—Engine manufacturer.
4—Propeller manufacturer.

4839. Propeller lubrication periods and lubricant specifications are determined by the

1—Federal Aviation Administration.
2—aircraft manufacturer.
3—propeller manufacturer.
4—American Petroleum Institute.

4840. Grease used in aircraft propellers reduces the frictional resistance of moving parts and is easily molded into any form under pressure. This statement defines

1—antifriction and plasticity characteristics of grease.
2—antifriction and chemical stability of grease.
3—antiwear properties and maximum cooling ability of grease.
4—viscosity and melting point of grease.

4841. What type of imbalance will cause a two–blade propeller to have a persistent tendency to come to rest in a horizontal position (with the blades parallel to the ground) while being checked on a propeller balancing beam?

1—Vertical.
2—Horizontal.
3—Dynamic.
4—Harmonic.

4842. What is the purpose of an arbor used in balancing a propeller?

1—To support the propeller on the balance knives.
2—To level the balance stand.
3—To indicate the weight to be added or removed.
4—To mark the propeller blades where weights are to be attached.

4843. If a blade of a particular metal propeller is shortened because of damage to the tip, the remaining blade(s) must be

1—ground down at the shank to balance the weight.
2—reset (blade angle) to compensate for the shortened blade.
3—returned to the manufacturer for alteration.
4—reduced to conform with the shortened blade.

4844. Which of the following statements is true regarding final balance of wood propellers?

1—If a separate metal hub is used, final balance should be accomplished with the hub installed in the propeller.
2—Solder should be applied to the light blade.
3—Final balance should be performed after the propeller has been installed.
4—The depth of the three holes in the tip of each blade should be varied to accomplish final horizontal balance.

4845. Apparent engine roughness is often a result of propeller unbalance. The effect of an unbalanced propeller will usually be

1—approximately the same at all speeds.
2—unnoticeable, except within the propeller's critical range.
3—greater at low RPM.
4—greater at high RPM.

4846. Which of the following is used to correct horizontal unbalance of a wood propeller?

1—Putty.
2—Brass screws.
3—Shellac.
4—Solder.

4847. A powerplant using a hydraulically controlled constant–speed propeller is operating within the propeller's constant speed range at a fixed throttle setting. If the tension of the propeller governor control spring (speeder spring) is reduced by movement of the cockpit propeller control, the propeller blade angle will

1—increase, engine manifold pressure will increase, and engine RPM will decrease.
2—decrease, engine manifold pressure will increase, and engine RPM will decrease.
3—decrease, engine manifold pressure will decrease, and engine RPM will increase.
4—increase, engine manifold pressure will decrease, and engine RPM will increase.

4848. Why is the pulley stop screw on a propeller governor adjustable?

1—To limit the maximum engine speed during takeoff.
2—To maintain the proper blade angle for cruising.
3—To limit the maximum propeller pitch for takeoff.
4—To maintain the most efficient engine speed for climbing.

4849. During engine operation at speeds lower than those for which the constant–speed propeller control can govern in the INCREASE RPM position, the propeller will

1—remain in the full LOW RPM position.
2—remain in full HIGH PITCH position.
3—maintain engine RPM in the normal manner until the HIGH PITCH stop is reached.
4—remain in the full LOW PITCH position.

4850. When engine power is increased, the constant–speed propeller tries to function so that it will

1—maintain the RPM, decrease the blade angle, and maintain a low angle of attack.
2—increase the RPM, decrease the blade angle, and maintain a low angle of attack.
3—maintain the RPM, increase the blade angle, and maintain a low angle of attack.
4—increase the RPM, increase the blade angle, and maintain a high angle of attack.

4851. The propeller governor controls the

1—oil to and from the pitch changing mechanism.
2—relief valve in the accumulator assemblies.
3—spring tension on the boost pump speeder spring.
4—linkage and counterweights from moving in and out.

4852. During the onspeed condition of a propeller, the

1—centrifugal force acting on the governor flyweights is greater than the tension of the speeder spring.
2—tension on the speeder spring is greater than the centrifugal force acting on the governor flyweights.
3—tension on the speeder spring is less than the centrifugal force acting on the governor flyweights.
4—centrifugal force of the governor flyweights is equal to the speeder spring force.

4853. What actuates the pilot valve in the governor of a constant–speed propeller?

1—Engine oil pressure.
2—Governor flyweights.
3—Propeller control lever.
4—Governor pump oil pressure.

4854. The operation of the pilot valve in the governor of a constant–speed propeller is controlled by the

1—blade counterweights.
2—booster pump oil pressure.
3—engine oil pressure.
4—centrifugal action of the flyweights.

4855. What action takes place when the cockpit control lever for a hydromatic, constant–speed propeller is actuated?

1—Tension of the speeder spring is changed.
2—The transfer valve changes position.
3—The governor booster pump pressure is varied.
4—The governor bypass valve is positioned to direct oil pressure to the propeller dome.

4856. What will happen to the propeller blade angle and the engine RPM if the tension on the propeller governor control spring (speeder spring) is increased?

1—Blade angle will increase and RPM will increase.
2—Blade angle will decrease and RPM will decrease.
3—Blade angle will increase and RPM will decrease.
4—Blade angle will decrease and RPM will increase.

4857. How is the speed of a hydromatic constant–speed propeller changed in flight?

1—By varying the output of the governor booster pump.
2—By advancing the throttle to a higher manifold pressure.
3—By changing the rotational speed of the pilot valve in the governor.
4—By changing the load tension against the flyweights in the governor.

4858. When the centrifugal force acting on the propeller governor counterweights overcomes the tension on the speeder spring, a propeller is in what speed condition?

1—Onspeed.
2—Underspeed.
3—In between condition.
4—Overspeed.

4859. What operational force causes the greatest stress on a propeller?

1—Aerodynamic twisting force.
2—Centrifugal force.
3—Thrust bending force.
4—Torque bending force.

4860. What operational force tends to increase propeller blade angle?

1—Centrifugal twisting force.
2—Aerodynamic twisting force.
3—Thrust bending force.
4—Torque bending force.

4861. How is a propeller controlled in a large aircraft with a turboprop installation?

1—Independently of the engine.
2—By varying the engine RPM except for feathering and reversing.
3—By varying the gear ratio between the propeller and the engine.
4—By the engine power lever.

4862. How does the aerodynamic twisting force affect operating propeller blades?

1—It tends to bend the blades opposite the direction of rotation.
2—It tends to turn the blades to a high blade angle.
3—It tends to bend the blades forward.
4—It tends to turn the blades to a low blade angle.

4863. Which of the following best describes the blade movement of a hydromatic propeller that is in the high RPM position when reversing action is begun?

1—Low pitch directly to reverse pitch.
2—No movement, because this type propeller cannot be placed in reverse pitch from the high RPM position.
3—Low pitch through high pitch to reverse pitch.
4—Low pitch through feather position to reverse pitch.

4864. Propellers exposed to salt spray should be cleaned with

1—a caustic solution.
2—steel wool.
3—fresh water.
4—soapy water.

4865. How can a steel propeller hub be tested for cracks?

1—By anodizing.
2—By magnafluxing.
3—By electrotesting.
4—By etching.

4866. Repairs of aluminum alloy adjustable pitch propellers are not permitted to be made on which of the following propeller blade areas?

1—Shank.
2—Leading edge.
3—Tip.
4—Trailing edge.

4867. Which of the following functions requires the use of a propeller blade station?

1—Measuring blade angle.
2—Installation and removal of propeller.
3—Indexing blades.
4—Propeller balancing.

4868. The propeller blade angle is defined as the acute angle between the airfoil section chord line (at the blade reference station) and which of the following?

1—The plane of rotation.
2—The relative wind.
3—The propeller thrust line.
4—The axis of blade rotation during pitch change.

4869. During which of the following conditions of flight will the blade pitch angle of a constant-speed propeller be the greatest?

1—Approach to landing.
2—Climb following takeoff.
3—High-speed, high-altitude cruising flight.
4—Takeoff from sea level.

4870. If a hydromatic propeller is feathered and then immediately unfeathers itself, a probable cause of the trouble is that the

1—governor is not cutting out in high pitch.
2—dome pressure relief valve is stuck in the CLOSED position.
3—distributor relief valve is stuck in the CLOSED position.
4—pressure cutout switch is stuck in the CLOSED position.

4871. The actual distance a propeller moves forward through the air during one revolution is known as the

1—effective pitch.
2—geometric pitch.
3—relative pitch.
4—resultant pitch.

4872. The pitch–changing mechanism of the hydromatic propeller is lubricated by

1—the pitch–changing oil.
2—using an approved–type grease in a grease gun at intervals prescribed by the propeller manufacturer.
3—applying an approved grease to the working surfaces.
4—thoroughly greasing, necessary only during propeller overhaul.

4873. What is the result of moving the throttle on a reciprocating engine when the propeller is in the constant–speed range with the engine developing cruise power?

1—Opening the throttle will cause an increase in blade angle.
2—Closing the throttle will cause an increase in blade angle.
3—The RPM will vary directly with any movement of the throttle.
4—Movement of the throttle will not affect the blade angle.

4874. Propeller blade stations are measured from the

1—index mark on the blade shank.
2—hub centerline.
3—blade base.
4—blade tip.

4875. The thrust produced by a rotating propeller is a result of

1—propeller slippage.
2—an area of low pressure behind the propeller blades.
3—an area of decreased pressure immediately in front of the propeller blades.
4—the angle of relative wind and rotational velocity of the propeller.

4876. Why is a constant–speed counterweight propeller normally placed in full HIGH PITCH position before the engine is stopped?

1—To prevent exposure and corrosion of the pitch changing mechanism.
2—To prevent hydraulic lock of the piston when the oil cools.
3—To prevent overheating of the engine during the next start.
4—To reduce engine temperatures more rapidly.

4877. The low pitch stop on a constant–speed propeller is usually set so that

1—the engine will turn at its rated takeoff RPM at sea level when the throttle is opened to allowable takeoff manifold pressure.
2—maximum allowable engine RPM cannot be exceeded with any combination of manifold pressure, altitude, or forward speed.
3—the limiting engine manifold pressure cannot be exceeded with any combination of throttle opening, altitude, or forward speed.
4—governing is permitted in cruising power descent from rated altitude.

4878. The angle–of–attack of a rotating propeller blade is measured between the blade chord or face and which of the following?

1—The plane of blade rotation.
2—Full low–pitch blade angle.
3—The relative air stream.
4—The geometric pitch angle required to produce the same thrust.

4879. The CTM (centrifugal twisting moment) of an operating propeller tends to

1—increase the pitch angle.
2—reduce the pitch angle.
3—bend the blades in the direction of rotation.
4—bend the blades rearward in the line of flight.

4880. Which of the following is identified as the cambered or curved side of a propeller blade, corresponding to the upper surface of a wing airfoil section?

1—Blade back.
2—Blade chord.
3—Blade leading edge.
4—Blade face.

4881. Which of the following best describes the blade movement of a full–feathering, constant–speed propeller that is in the LOW RPM position when the feathering action is begun?

1—High pitch through low pitch to feather position.
2—Low pitch directly to feather position.
3—High pitch directly to feather position.
4—Low pitch through high pitch to feather position.

4882. The holding coil on a hydromatic propeller feathering button switch holds a solenoid relay closed that applies power to the propeller

1—governor.
2—synchronizer.
3—dome feathering mechanism.
4—feathering motor pump.

4883. What is the primary purpose of the metal tipping which covers the blade tips and extends along the leading edge of each wood propeller blade?

1—To increase the lateral strength of the blade.
2—To prevent impact damage to the tip and leading edge of the blade.
3—To increase the longitudinal strength of the blade.
4—To provide a true airfoil the entire length of the blade.

4884. Blade angle is an angle formed by a line perpendicular to the crankshaft and a line formed by the

1—relative wind.
2—apparent wind.
3—chord of the blade.
4—blade face.

4885. Propeller blade station numbers increase from

1—hub to tip.
2—tip to hub.
3—leading edge to aft edge.
4—aft edge to leading edge.

4886. The aerodynamic force acting on a rotating propeller blade operating at a normal pitch angle tends to

1—reduce the pitch angle.
2—increase the pitch angle.
3—bend the blades rearward in the line of flight.
4—bend the blades in the direction of rotation.

4887. Hydromatic propeller pitch change gear preload may be adjusted by

1—moving the vernier preload lockplate clockwise to increase preload and anticlockwise to reduce preload.
2—adjusting the stop plate that limits the movement of the movable cam within the stationary cam.
3—varying the spider shim thickness used between the spider shim plate and the blade bushing face.
4—varying the thickness of the shims between the fixed cam base and the dome–barrel shelf.

4888. Which of the following forces or combination of forces operates to move the blades of a constant–speed counterweight type propeller to the HIGH PITCH position?

1—Engine oil pressure acting on the propeller piston–cylinder arrangement.
2—Engine oil pressure acting on the propeller piston–cylinder arrangement and centrifugal force acting on the counterweights.
3—Centrifugal force acting on the counterweights.
4—Prop governor oil pressure acting on the propeller piston–cylinder arrangement.

4889. The distributor valve assembly of a hydromatic propeller changes position and reverses the oil passages to the propeller only when the propeller is

1—being unfeathered.
2—being feathered.
3—in the overspeed condition.
4—in the underspeed condition.

4890. Which of the following best describes the blade movement of a feathering propeller that is in the HIGH RPM position when the feathering action is begun?

1—High pitch through low pitch to feather position.
2—Low pitch through reverse pitch to feather position.
3—No movement, because a feathering propeller cannot be feathered from the HIGH RPM.
4—Low pitch through high pitch to feather position.

4891. Which of the following is normally recommended for the routine care of solid aluminum propeller blades?

1—Prevent any type of petroleum product such as fuel and oil from remaining in contact with the blades by frequently cleaning with soap and water.
2—Clean the blades with gasoline or other volatile cleaner and wipe dry with a clean, soft cloth.
3—Wash the blades with soap and water, rinse with clear water, dry thoroughly, and apply a thin coat of clean engine oil.
4—Clean the blades with a commercial caustic compound and rinse with clear water.

4892. The blade angle of a fixed–pitch propeller

1—is greatest at the tip.
2—is constant from the hub to the tip.
3—is smallest at the tip.
4—increases in proportion to the distance each section is from the hub.

4893. What determines the amount which an aluminum alloy propeller blade can be bent in face alignment and be repairable by cold straightening?

1—The thickness of the blade section at which the bend is located.
2—The chord length of the blade section at which the bend is located.
3—The linear distance from the blade tip at which the bend is located.
4—The linear distance from the propeller centerline at which the bend is located.

4894. During operational check of an aircraft using hydromatic full–feathering propellers, the following observations are made:

The feather button, after being pushed, remains depressed until the feather cycle is complete, then opens.

When unfeathering, it is necessary to manually hold the button down until unfeathering is accomplished.

1—Both feather cycle and unfeather cycle are functioning properly.
2—Both feather and unfeather cycles indicate malfunctions.
3—The feather cycle indicates a malfunction, but the unfeather cycle is correct.
4—The feather cycle is correct. The unfeather cycle indicates a malfunction.

4895. The etching process is used during propeller overhaul to

1—detect blade defects.
2—identify unairworthy components.
3—identify the blades.
4—indicate the overhauling agency name and certificate number.

4896. What controls the constant–speed range of a constant–speed propeller?

1—Engine RPM.
2—Angle of climb and descent with accompanying changes in airspeed.
3—Number of blades.
4—The mechanical limits in the propeller pitch range.

4897. For takeoff, a constant–speed propeller is normally set in the

1—HIGH PITCH, high RPM position.
2—LOW PITCH, low RPM position.
3—HIGH PITCH, low RPM position.
4—LOW PITCH, high RPM position.

4898. Where are the high and low pitch stops of a constant–speed or two–position counterweight propeller located?

1—On the face of the pitch change thrust plate.
2—In the hub and blade assembly.
3—In the counterweight assembly.
4—In the dome assembly.

4899. Which of the following statements about constant–speed counterweight propellers is also true when referring to two–position counterweight propellers?

1—Blade angle changes are accomplished by the use of two forces, one hydraulic and the other centrifugal.
2—A range of blade angle travel of either 15° or 20° is available.
3—Since an infinite number of blade angle positions are possible during flight, propeller efficiency is greatly improved.
4—The pilot selects the RPM and the propeller changes pitch to maintain the selected RPM.

4900. Most engine–propeller combinations have one or more critical ranges within which continuous operation is not permitted. Critical ranges are established to avoid

1—severe propeller vibration.
2—severe turbulence within the slipstream.
3—low or negative thrust conditions.
4—inefficient propeller pitch angles.

4901. Which of the following defects is cause for rejection of wood propellers?

1—Solder missing from screw heads securing metal tipping.
2—An oversize hub or bolt hole, or elongated bolt holes.
3—No moisture drain holes in metal tipping.
4—No protective coating on propeller.

4902. The centrifugal load of the rotating blades of a counterweight or hydromatic propeller is transferred to the

1—barrel.
2—spider.
3—thrust washer.
4—barrel support assembly.

4903. Why is it important that nicks in the leading edge of aluminum alloy blades be removed as soon as possible?

1—To localize vibratory stress.
2—Horizontal balance purposes.
3—To improve the aerodynamic characteristics of the blades.
4—To eliminate the condition where fatigue cracks can start.

4904. Major repairs to aluminum alloy propellers and blades may be done by

1—a powerplant mechanic working for a certificated A & P mechanic.
2—any propeller manufacturer.
3—an appropriately rated repair station or the manufacturer.
4—a repairman, regardless of where he/she works.

4905. The primary purpose of a cuff on a propeller is to

1—distribute anti–icing fluid.
2—strengthen the propeller.
3—gain a smoother airflow thereby reducing drag.
4—increase the flow of cooling air to the engine nacelle.

4906. The purpose of a three–way propeller valve is to

1—direct oil from the engine oil system to the propeller cylinder.
2—permit the governor to maintain on–speed condition.
3—direct oil from the engine through the governor to the propeller.
4—permit constant–speed operation of the propeller.

4907. The primary purpose of a propeller is to

1—create lift on the fixed airfoils of an aircraft.
2—build up enough slipstream to support the airfoils.
3—change engine horsepower to thrust.
4—provide static and dynamic stability of an aircraft in flight.

4908. A constant–speed propeller provides maximum efficiency by

1—increasing blade pitch as the aircraft speed decreases.
2—adjusting blade angle for most conditions encountered in flight.
3—reducing turbulence near the blade tips.
4—increasing the lift coefficient of the blade.

4909. The centrifugal twisting force acting on a propeller blade is

1—greater than the aerodynamic twisting force and tends to move the blade to a higher angle.
2—less than the aerodynamic twisting force and tends to move the blade to a lower angle.
3—less than the aerodynamic twisting force and tends to move the blade to a higher angle.
4—greater than the aerodynamic twisting force and tends to move the blade to a lower angle.

4910. Geometric pitch of a propeller is defined as the

1—effective pitch minus slippage.
2—effective pitch plus slippage.
3—angle between the blade chord and the plane of rotation.
4—angle between the blade face and the plane of rotation.

4911. Propeller blade angle is the angle between the

1—chord of the blade and the relative wind.
2—relative wind and the rotational plane of the propeller.
3—chord of the blade and the rotational plane of the propeller.
4—geometric pitch and the effective pitch.

4912. What operational force causes propeller blade tips to lag in the opposite direction of rotation?

1—Thrust–bending force.
2—Aerodynamic–twisting force.
3—Centrifugal–twisting force.
4—Torque–bending force.

4913. What operational force tends to bend the propeller blades forward at the tip?

1—Torque–bending force.
2—Aerodynamic–twisting force.
3—Centrifugal–twisting force.
4—Thrust–bending force.

4914. What are the rotational speed and blade pitch angle requirements of a constant–speed propeller during takeoff?

1—Low–speed and low–pitch angle.
2—Low–speed and high–pitch angle.
3—High–speed and low–pitch angle.
4—High–speed and high–pitch angle.

4915. Holes are drilled in the metal tips of fixed–pitch wood propellers to

1—decrease the weight of the propeller.
2—minimize splitting along the grain of the wood.
3—balance the propeller.
4—equalize the moisture content within the blades.

4916. When running–up an engine and testing a newly installed hydromatic propeller, it is necessary to exercise the propeller by moving the governor control through its entire travel several times to

1—seat the blades fully against the low pitch stop.
2—check the minimum RPM setting of the governor.
3—free the dome of any entrapped air.
4—test the maximum RPM setting of the governor.

4917. Which of the following occurs to cause front cone bottoming during propeller installation?

1—The front cone becomes bottomed in the front propeller hub cone seat before the rear propeller hub cone seat has engaged the rear cone.
2—The front cone becomes bottomed in the front propeller hub cone seat before the retaining nut has engaged sufficient threads to be safetied properly.
3—The front cone enters the front propeller hub cone seat at an angle causing the propeller retaining nut to appear tight when it is only partially tightened.
4—The front cone contacts the ends of the shaft splines, preventing the front and rear cones from being tightened against the cone seats in the propeller hub.

4918. A damaged piston–to–dome seal in a hydromatic propeller will most likely be indicated by which of the following?

1—Oil deposits on the propeller hub and blades.
2—A heavy oil deposit on the engine nose case and cowling with little, if any, oil on the blades.
3—Oil deposits on the blades and the outer portions of the engine cowling with no oil on the propeller hub.
4—Sluggish operation of the pitch change mechanism.

4919. During the installation of a hydromatic propeller, the blades should be in what position when the dome is to be installed?

1—Reverse pitch.
2—Full–feathered.
3—Low pitch.
4—High pitch.

4920. What is indicated when the front cone bottoms while installing a propeller?

1—Propeller retaining nut torque is correct.
2—Propeller–dome combination is incorrect.
3—Blade angles are incorrect.
4—Rear cone should be moved forward.

4921. How is the oil pressure delivery on a hydromatic propeller normally stopped after the blades have reached their full feathered position?

1—Pulling out the feathering push button.
2—Electric cutout pressure switch.
3—High–angle stop ring in the base of the fixed cam.
4—Stop lugs in the teeth of the rotating cam.

4922. Incorrect pitch change gear preload on a hydromatic propeller will cause

1—insufficient drag to be exerted on the blades, resulting in erratic tracking.
2—excessive clearances between the blade and spider.
3—excessive binding or backlash of the gear teeth.
4—the propeller to become loose on the shaft after operation.

4923. The primary purpose of the front and rear cones for propellers that are installed on splined shafts is

1—to position the propeller hub on the splined shaft.
2—to prevent metal–to–metal contact between the propeller and the splined shaft.
3—to reduce stresses between the splines of the propeller and the splines of the shaft.
4—to balance the propeller aerodynamically.

4924. Which of the following statements concerning the installation of a new fixed–pitch wood propeller is true?

1—If the hub flange is integral with the crankshaft, final track should be made before the propeller is installed.
2—If a separate metal hub is used, final track should be accomplished prior to installing the hub in the propeller.
3—NAS close–tolerance bolts should be used to install the propeller.
4—Inspect the bolts for tightness after the first flight and again after the first 25 hours of flying.

4925. If the propeller cones or the hub cone seats show evidence of galling and wear, the most likely cause is

1—the cones and cone seats were not properly lubricated during previous operation.
2—the pitch change stops were located incorrectly, causing the cone seats to act as the high pitch stop.
3—the propeller retaining nut was not tight enough during previous operation.
4—the front cone was not fully bottomed against the crankshaft splines during installation.

4926. On aircraft equipped with hydraulically operated constant–speed propellers, all ignition and magneto checking is done with the propeller in which position?

1—High RPM.
2—Normal cruising range.
3—Low RPM.
4—High pitch range.

4927. Oil leakage around the rear cone of a hydromatic propeller usually indicates a defective

1—piston gasket.
2—blade–barrel packing.
3—spider–shaft oil seal.
4—dome–barrel oil seal.

4928. Maximum taper contact between crankshaft and propeller hub is determined by using

1—a telescoping gauge.
2—bearing blue color transfer.
3—a micrometer.
4—a surface gauge.

4929. Propeller blade tracking is the process of determining

1—the plane of rotation of the propeller with respect to the aircraft longitudinal axis.
2—that each blade will have the same angle of attack to prevent vibration.
3—that the blade angles are within the specified tolerance of each other.
4—the positions of the tips of the propeller blades relative to each other.

4930. What is the basic purpose of the three small holes (No. 60 drill) in the tipping of wood propeller blades?

1—To provide a means for inserting balancing shot when necessary.
2—To provide a means for periodically impregnating the blade with preservation materials.
3—To allow the moisture which may collect between the tipping and the wood to escape (vent the tipping).
4—To allow the moisture content of each blade to equalize.

4931. A fixed–pitch wooden propeller that has been properly installed and the attachment bolts properly torqued, exceeds the out–of–track allowance by 1/16 inch. The excessive out–of–track condition may be corrected by

1—slightly overtightening the attachment bolts adjacent to the most forward blade.
2—discarding the propeller since out–of–track conditions cannot be corrected.
3—re–profiling the blades to correct for unequal aerodynamic forces.
4—placing paper shims between the inner flange and the propeller.

4932. Manually feathering a hydromechanical propeller means to

1—block governor oil pressure to the cylinder of the propeller.
2—port governor oil pressure to the cylinder of the propeller.
3—port governor oil pressure from the cylinder of the propeller.
4—block governor oil pressure in the cylinder of the propeller.

4933. In what position is the constant–speed propeller control placed to check the magnetos?

1—Full decrease, low propeller blade pitch angle.
2—Full increase, high propeller blade pitch angle.
3—Full increase, low propeller blade pitch angle.
4—Full decrease, high propeller blade pitch angle.

4934. The propeller–feathering pump is shut off

1—15 seconds after depressing the feather button switch.
2—by a micro switch in the propeller governor.
3—by an oil pressure switch.
4—when the propeller piston activates the limit switch.

4001. ANSWER #2
Ball bearings provide less friction than any other type. A ball bearing consists of an inner race and an outer race, a set of polished steel balls, and a ball retainer. Because the steel balls offer less surface contact than do the rollers of the roller bearing, there is less rolling friction.
REFERENCE: AIRCRAFT POWERPLANTS, Page 49

4002. ANSWER #4
The question identifies a propeller gear reduction ratio, which means the propeller will be turning slower than the engine. In answer #4, the engine would turn three times for each two revolutions of the propeller (3:2). This is the least gear reduction of the four choices, so this would produce the highest propeller RPM for a given engine speed.
REFERENCE: ITP POWERPLANT, Chapter 1, Page 38,
        AC65-12A, Page 26

4003. ANSWER #3
One of the purposes of oil, in relation to engine bearings, is to prevent metal to metal contact. If a bearing is in the process of failing, metal to metal contact will be occuring. The friction which accompanies this contact will generate a great deal of heat and will cause high oil temperature.
REFERENCE: ITP POWERPLANT, Chapter 8, Page 32,
        AC65-12A, Page 457

4004. ANSWER #3
A large propeller, moving a large mass of air by turning at a low RPM, is much more efficient than a small propeller moving the same mass of air by turning at a high RPM. By using propeller reduction gearing, the RPM of the propeller can be kept at an efficient level while allowing the engine to operate at a higher, more efficient RPM.
REFERENCE: ITP POWERPLANT, Chapter 1, Page 37
        AC65-12A, Page 25

4005. ANSWER #4
Volumetric efficiency is a comparison of the volume of fuel/air charge (corrected for temperature and pressure) inducted into the cylinders to the total piston displacement of the engine. Excessive cylinder head temperature will decrease volumetric efficiency for two reasons. It will inhibit the flow of fuel/air charge into the cylinder, and it will decrease the density of the charge.
REFERENCE: ITP POWERPLANT, Chapter 1, Page 13
        AC65-12A, Page 38

4006. ANSWER #4
Special deep-groove ball bearings are used in aircraft radial engines to transmit propeller thrust to the engine nose section.
REFERENCE: AC65-12A, Page 25

4007. ANSWER #2
Master rod bearings on radial engines are generally plain bearings.
REFERENCE: AC65-12A, Page 25

4008. ANSWER #2
Indicated horsepower is the power developed in the combustion chambers without reference to friction losses within the engine.
REFERENCE: ITP POWERPLANT, Chapter 1, Page 11
        AC65-12A, Page 33

4009. ANSWER #4
The piston displacement of an engine is equal to the displacement or volume of one cylinder, multiplied by the number of cylinders. The volume of a cylinder, is equal to the area of the piston (π times the radius squared) multiplied by the stroke of the piston. For this engine, the displacement is 1,283 cubic inches.
REFERENCE: AC65-12A, Pages 30 and 31

4010. ANSWER #3
The four-stroke cycle officially begins when the piston starts moving down in the cylinder on the intake stroke. When the piston reaches bottom center and starts moving up in the cylinder, the compression stroke is occurring. Near the top of the compression stroke, the spark plug fires and the power stroke begins. As the piston approaches bottom center, the exhaust valve opens and the exhaust stroke begins.
REFERENCE: ITP POWERPLANT, Chapter 1, Page 8
             AC65-12A, Pages 29 and 30

4011. ANSWER #4
As the piston reaches the top of the compression stroke, the fuel/air charge is fired by means of an electric spark. The time of ignition will vary from 20 to 35 degrees before top dead center, depending upon the requirements of the specific engine, to ensure complete combustion of the charge by the time the piston is slightly past the top dead center position.
REFERENCE: AC65-12A, Page 29

4012. ANSWER #3
If the clearance between the rocker arm and the valve tip is decreased on a piston engine, the following will occur:
1.   the valve will open early and close late
2.   the valve will open more (the height of valve opening will be more)
3.   the valve will be open longer
REFERENCE: AC65-12A, Page 467

4013. ANSWER #2
A crankshaft is statically balanced when the weight of the entire assembly of crankpins, crank cheeks, and counterweights is balanced around the axis of rotation. The counterweights are varied in order to gain static balance.
REFERENCE: AC65-12A, Page 12

4014. ANSWER #4
For a reciprocating engine to operate properly, each valve must open at the proper time, stay open for the required length of time, and close at the proper time. Intake valves are opened just before the piston reaches top dead center, and exhaust valves remain open after top dead center. At a particular instant, therefore, both valves are open at the same time (end of the exhaust stroke and beginning of the intake stroke).
REFERENCE: ITP POWERPLANT, Chapter 1, Pages 9 and 10
             AC65-12A, Page 21

4015. ANSWER #3
It is possible, when using a DC continuity tester, to force enough current through the primary coil to produce a magnetic field that will weaken the permanent magnet. The main reason for disconnecting the primary circuit between the coil and the breaker points when using the tester is to make it possible to tell when the points open. If the primary circuit is connected, current flow through the primary winding of the coil will keep the light illuminated even when the points open.
REFERENCE: THE AVIATION MECHANIC'S MANUAL, Page 338
             AIRCRAFT PROPULSION POWERPLANTS, Page 298

4016. ANSWER #1
Several aircraft recip engines now use cam ground pistons to compensate for the greater expansion parallel to the piston pin during engine operation. The diameter of these pistons measures several thousandths of an inch larger at an angle to the piston pin hole than parallel to the pin hole.
REFERENCE: AC65-12A, Page 420

4017. ANSWER #3
This question indicates that the intake valve closes 45 degrees after bottom dead center, or *135 degrees before top dead center*. The exhaust valve opens 70 degrees before bottom dead center, or *110 degrees after top dead center*. 135 degrees plus 110 degrees, or 245 degrees, is the number of degrees both valves are seated.
REFERENCE: ITP POWERPLANT, Chapter 1, Pages 9 and 10

4018. ANSWER #1
In a zero-lash hydraulic valve lifter, oil pressure pumps up the lifter to remove all clearance between the rocker arm and the valve stem. This condition being present in an engine is normal.
REFERENCE: ITP POWERPLANT, Chapter 1, Page 41
             AC65-12A, Page 23

4019. ANSWER #3
When setting up the ignition timing on an engine, a timing disk can be attached to the engine to measure the crankshaft rotation in degrees.
REFERENCE: AC65-12A, Page 198

4020. ANSWER #3
As the piston leaves the top-center position, it accelerates and attains its maximum speed when the connecting rod and crank throw are at right angles (90 degrees).
REFERENCE: ITP POWERPLANT, Chapter 1, Page 9
             AIRCRAFT POWERPLANTS, Page 54

4021. ANSWER #1
The cylinder barrel of recip engines is made of a steel alloy forging with the inner surface hardened to resist wear of the piston and the piston rings which bear against it. This hardening is usually done by exposing the steel to ammonia or cyanide gas while the steel is very hot. The steel soaks up nitrogen from the gas which forms iron nitrides on the exposed surface. As a result of this process, the metal is said to be nitrided.
REFERENCE: ITP POWERPLANT, Chapter 1, Page 29
             AC65-12A, Page 18

4022. ANSWER #1
The intake valve of a recip engine is timed to close about 50 to 75 degrees past bottom dead center on the compression stroke, depending upon the specific engine, to allow the momentum of the incoming gases to charge the cylinder more completely.
REFERENCE: ITP POWERPLANT, Chapter 1, Page 8
             AC65-12A, Page 29

4023. ANSWER #2
Intake and exhaust valves can be checked for stretch by one of two methods. One involves checking the diameter of the valve stem near the neck of the valve with a micrometer. If the diameter is smaller than normal, it indicates the valve has been stretched. The other method involves checking the valve with a radius or contour gage. The contour is designed to fit along the underside of the valve head. If the contour of the gage and that of the valve do not match, it indicates that the valve has been stretched.
REFERENCE: AC65-12A, Page 418

4024. ANSWER #3
In a recip engine, the fuel/air charge is fired by means of an electric spark as the piston approaches top dead center on the compression stroke. The time of ignition will vary from 20 to 35 degrees before top dead center, depending upon the requirements of the specific engine, to ensure complete combustion of the charge by the time the piston is slightly past the top dead center position.
REFERENCE: ITP POWERPLANT, Chapter 1, Page 8
             AC65-12A, Page 29

4025. ANSWER #2
The following number of degrees of travel occur between the spark plug firing and the intake valve opening:
SPARK PLUG FIRES . . . . 28 degrees before top center
PISTON TRAVEL . . . . 180 degrees on the power stroke (top center to bottom center)
INTAKE VALVE OPENS . . . . 165 degrees after bottom center
The total number of degrees of travel is 373.
REFERENCE: ITP POWERPLANT, Chapter 1, Page 10

4026. ANSWER #4
The stems of some valves have a narrow groove below the lock-ring groove for the installation of safety circlets or spring rings. The circlets are designed to prevent the valves from falling into the combustion chambers if the tip should break during engine operation and on the occasion of valve disassembly and assembly.
REFERENCE: AIRCRAFT POWERPLANTS, Page 52

4027. ANSWER #2
There are many reasons why proper valve clearances are of vital importance to satisfactory and stable engine operation. For example, when the engine is operating, valve clearances establish valve timing. If the clearance is too great, the valves will open late and close early, which will severely affect engine performance.
REFERENCE: AC65-12A, Page 467

4028. ANSWER #2
Some aircraft engines incorporate hydraulic tappets which automatically keep the valve clearance at zero, eliminating the necessity for any valve clearance adjustment mechanism.
REFERENCE: ITP POWERPLANT, Chapter 1, Page 41
AC65-12A, Page 23

4029. ANSWER #4
The intake valve of a recip engine is timed to close about 50 to 75 degrees past bottom dead center on the compression stroke, depending upon the specific engine. This is done to allow the momentum of the incoming gases to charge the cylinder more completely.
REFERENCE: ITP POWERPLANT, Chapter 1, Pages 8 and 10
AC65-12A, Page 29

4030. ANSWER #4
The maximum compression ratio of an engine is limited by the detonation characteristics of the fuel used. Detonation occurs when 75 to 80 percent of the fuel/air mixture burns normally and then the remainder burns with explosive rapidity. The maximum compression ratio of an aircraft engine is also limited by the design limitations of the engine, the availability of high-octane fuel, and the degree of supercharging.
REFERENCE: ITP POWERPLANT, Chapter 1, Pages 13 and 14
AIRCRAFT POWERPLANTS, Page 12
AC65-12A, Page 31

4031. ANSWER #2
The piston pin used in modern aircraft engines is the full-floating type, so called because the pin is free to rotate in both the piston and in the connecting rod piston-pin bearing.
REFERENCE: ITP POWERPLANT, Chapter 1, Pages 33 and 34
AC65-12A, Pages 15 and 16

4032. ANSWER #4
For a reciprocating engine to operate properly, each valve must open at the proper time, stay open for the required length of time, and close at the proper time. Intake valves are opened just before the piston reaches top dead center, and exhaust valves remain open after dead center. At a particular instant, therefore, both valves are open at the same time. This valve-overlap permits better volumetric efficiency and lowers the cylinder operating temperature.
REFERENCE: ITP POWERPLANT, Chapter 1, Page 9
AC65-12A, Page 21

4033. ANSWER #2
When an engine is hot, the clearance between the rocker arm and the stem of the valve is greater than it is when the engine is cold. If the engine is cold and the valves are set using the hot clearance, the setting will be too great. This will cause the valves to open late and close early.
REFERENCE: AC65-12A, Page 467

4034. ANSWER #3
Each valve is closed by two or three helical-coiled springs. If a single spring were used, it would vibrate or surge at certain speeds.
REFERENCE: ITP POWERPLANT, Chapter 1, Page 31
AC65-12A, Page 24

4035. ANSWER #3
Modern aircraft engines are constructed of such durable materials that top overhaul has largely been eliminated. Top overhaul means overhaul of those parts "on top" of the crankcase without completely dismantling the engine. The actual top overhaul consists of reconditioning the cylinder, piston, and valve-operating mechanism, and replacing the valve guides and piston rings, if needed.
REFERENCE: ITP POWERPLANT, Chapter 1, Page 59
AC65-12A, Page 411

4036. ANSWER #1
Airframes are built as light as modern technology will allow and with lightness there is a low resistance to vibration. For this reason, the aircraft engine must operate as smoothly as possible. The more cylinders the engine has, the more the power impulses will overlap and the smoother the engine will operate.
REFERENCE: ITP POWERPLANT, Chapter 1, Page 19

4055. ANSWER #2
Metal particles on the engine oil screens or the magnetic sump plugs are generally an indication of partial internal failure of the engine. However, due to the construction of aircraft oil systems, it is possible that metal particles may have collected in the oil system sludge at the time of a previous engine failure. At any rate, the cause of the particles should be determined before the engine is returned to service.
REFERENCE: AC65-12A, Page 360

4056. ANSWER #2
The most likely cause of oil pressure fluctuating between zero and normal oil pressure is a low oil supply. So long as the oil is being picked up by the pump, the oil pressure will be normal. The low oil supply, however, will cause momentary losses of oil pick-up. During this period of time, the oil pressure will drop to zero.
REFERENCE: ITP POWERPLANT, Chapter 1, Page 77; Chapter 8, Pages 30 and 31
                      AC65-12A, Page 301

4057. ANSWER #3
When adjusting the valves on an engine with a floating cam ring, the valves must be depressed and released simultaneously and smoothly. The valves must be unloaded to remove the spring tension from the side positions on the cam and thus permit the cam to slide away from the valves to be adjusted until it contacts the cam bearing. This locates the cam in a definite position and prevents cam shift from introducing errors in the clearance.
REFERENCE: AC65-12A, Page 468

4058. ANSWER #3
On engines which use overhead valves, valve clearance decreases with a drop in temperature; therefore, insufficient clearance may cause the valve to hold open when extremely cold temperatures are encountered. If the clearance is much too small, the valves may not seat properly during the start and warmup phase of operation.
REFERENCE: AC65-12A, Page 467

4059. ANSWER #2
When there is too much valve clearance, the valves will not open as wide or remain open as long as they should. This reduces the overlap period.
REFERENCE: AC65-12A, Page 450

4060. ANSWER #4
Valve-overlap permits better volumetric efficiency and lowers the cylinder operating temperature.
REFERENCE: ITP POWERPLANT, Chapter 1, Page 9
                      AC65-12A, Page 21

4061. ANSWER #4
Most aircraft engine oil systems, whether dry-sump or wet-sump, have an oil pressure regulating valve. This valve maintains the oil pressure at a certain value once that pressure is obtained. It is normal for oil pressure to be less at idle than at cruise RPM, because there is less oil flowing and therefore less resistance to flow. At idle RPM, there is not enough oil flow to build a pressure equal to that of the regulating valve's setting.
REFERENCE: AC65-12A, Page 300

4062. ANSWER #4
In a four-stroke cycle engine, each cylinder in the engine fires once in every two revolutions of the crankshaft. For the cylinder to fire 200 times in a minute, the crankshaft must rotate 400 times a minute.
REFERENCE: ITP POWERPLANT, Chapter 1, Page 10 (Fig. 2A-7)
                      AC65-12A, Page 28

4063. ANSWER #2
Crankshaft run-out is checked when an engine is being overhauled and when the engine is subjected to a sudden stoppage.
REFERENCE: AC65-12A, Page 428
                      AC43.13-1A, Page 267

4064. ANSWER #4
Whenever a radial engine remains shut down for any length of time beyond a few minutes, oil or fuel may drain into the combustion chambers of the lower cylinders or accumulate in the lower intake pipes. These fluids can cause a liquid lock, which will damage the engine if it is rotated with the starter. To check for a liquid lock, the propeller should be turned by hand in the normal direction of rotation three to four revolutions.
REFERENCE: ITP GENERAL, Chapter 7, Page 25
                        AC65-12A, Page 457
                        AC65-9A, Page 489

4065. ANSWER #4
A four lobe cam ring used on a nine cylinder engine turns at 1/8 crankshaft speed. A one lobe cam ring would rotate at 1/2 crankshaft speed, a two lobe cam ring at 1/4 crankshaft speed, and a three lobe cam ring at 1/6 crankshaft speed.
REFERENCE: ITP POWERPLANT, Chapter 1, Page 45
                        AC65-12A, Page 23

4066. ANSWER #1
The cold cylinder check determines the operating characteristics of each cylinder of an air-cooled engine. The tendency for any cylinder or cylinders to be cold or to be only slightly warm indicates lack of combustion or incomplete combustion within the cylinder. If an engine misses in both the right and left positions of the magneto switch, a check for a cold cylinder might indicate where the problem lies.
REFERENCE: AC65-12A, Page 470

4067. ANSWER #2
Valve blow-by is indicated by a hissing or whistle when pulling the propeller through prior to starting the engine, when turning the engine with the starter, or when running the engine at slow speeds. Blow-by past the exhaust valve can be heard at the exhaust stack.
REFERENCE: AC65-12A, Page 458

4068. ANSWER #2
As the oil in an operating engine heats up, it thins out. The thinner oil offers less resistance to flow, so it is normal for the oil pressure to be less when the engine is heated up than when the engine is cold. This would be especially true at idle and low speeds.
REFERENCE: ITP POWERPLANT, Chapter 8, Page 10
                        AC65-12A, Page 437

4069. ANSWER #2
If the oil dilution valve in an engine is leaking, it means that it is leaking fuel into the oil. This will thin the oil out, which will cause the oil pressure to be low. Because the fuel thinned oil cannot transfer heat as readily as normal oil, the oil temperature will be high.
REFERENCE: AC65-12A, Page 301

4070. ANSWER #4
The fuel/air mixture which will result in the highest engine temperature varies according to the engine's power setting. At cruise power settings, a mixture of 15 parts air to 1 part fuel (.067) will produce the highest temperature. At high power settings, lean mixtures produce the highest temperature. A mixture leaner than .060, at high power settings, will produce very high temperatures.
REFERENCE: AC65-12A, Pages 111 to 113

4071. ANSWER #2
Before removing a cylinder from the engine, the piston should be at top dead center on the compression stroke.
REFERENCE: AC65-12A, Page 462

4072. ANSWER #1
As a radial engine warms up, the clearance between the rocker arm and the valve stem increases.
REFERENCE: AC65-12A, Page 467

4073. ANSWER #3
The firing order of a nine cylinder radial engine is 1, 3, 5, 7, 9, 2, 4, 6, 8.
REFERENCE: ITP POWERPLANT, Chapter 1, Page 24
                        AC65-12A, Pages 19 and 22

4074. ANSWER #3
Operating flexibility is the ability of an engine to run smoothly and give desired performance at all speeds from idling to full-power output.
REFERENCE: ITP POWERPLANT, Chapter 1, Page 19
            AC65-12A, Page 4

4075. ANSWER #2
Generally, standard aircraft cylinder oversizes are 0.010 inch, 0.015 inch, 0.020 inch, or 0.030 inch. Unlike automobile engines which may be re-bored as much as 0.100 inch, aircraft cylinders have relatively thin walls and may have a nitrided surface, which must not be ground away.
REFERENCE: AC65-12A, Page 427

4076. ANSWER #4
The magneto safety check is conducted with the propeller in the high RPM position, at approximately 1,000 RPM. The ignition switch is moved from the "both" to the "right" position, and then the "both" to the "left" position. While switching from "both" to a single magneto position, a slight but noticeable drop in RPM should occur.
REFERENCE: ITP POWERPLANT, Chapter 3, Page 24
            AC65-12A, Pages 210, 211, and 437

4077. ANSWER #4
The most likely cause of a rough running engine which has normal magneto drop and high manifold pressure is a dead cylinder. A leak in the intake manifold could also cause a rough running engine, but the manifold pressure would not necessarily be high (not on a supercharged engine), nor higher than normal at all RPM's.
REFERENCE: AC65-12A, Page 438

4078. ANSWER #1
If the valve guides of an engine are worn, there will be too much clearance between the guide and the valve stem. This will allow oil to get by and enter the cylinder, causing high oil consumption.
REFERENCE: AC65-12A, Page 416

4079. ANSWER #2
When the piston in cylinder #1 is rotated 260 degrees past top dead center of the compression stroke, cylinder #3 will be 220 degrees past top dead center of the compression stroke. This puts cylinder #3 on the exhaust stroke, which would explain why the cylinder won't hold any pressure.
REFERENCE: AC65-12A, Page 22

4080. ANSWER #4
With an increase in manifold pressure, there will be an increase in power produced by an engine for a given RPM. With an increase in power, there will be an increase in the load on the master rod bearing.
REFERENCE: AIRCRAFT POWERPLANTS, Page 173

4081. ANSWER #1
The carburetor heat control must be rigged with springback to ensure that the air valve is completely open or closed when the control is in the "heat off" or "heat on" position. With springback, the valve is completely closed or completely open before the control reaches the limit of its travel.
REFERENCE: AIRCRAFT POWERPLANTS, Page 138

4082. ANSWER #1
Because of the reduced density of the air at altitude, as an aircraft ascends there is less air available to mix with the fuel being combusted by the engine. Unless there is an automatic or manual device to correct the problem, such as an automatic mixture control, the fuel/air mixture will become richer as altitude increases.
REFERENCE: AC65-12A, Page 111

4083. ANSWER #1
Because water vapor is a non-combustible gas which displaces oxygen in the air, its presence in the atmosphere is a total loss as far as contributing to engine power output. The mixture of water vapor and air is drawn through the carburetor and fuel is metered into it as though it were all air. This mixture of water vapor, air, and fuel enters the combustion chamber where it is ignited. Since the water vapor will not burn, the effective fuel/air ratio is enriched and the engine operates as though it were on an excessively rich mixture.
REFERENCE: ITP POWERPLANT, Chapter 6, Page 4
            AC65-9A, Page 239

4084. ANSWER #2
Unless detonation is heavy, there is no cockpit evidence of its presence. Light to medium detonation does not cause noticeable roughness, temperature increase, or loss of power.
REFERENCE: AC65-12A, Page 444

4085. ANSWER #3
Before a new engine is flight-tested, it must undergo a thorough ground check. Before this ground check can be made, several operations are usually performed on the engine. To prevent failure of the engine bearings during the initial start, the engine should be pre-oiled. When an engine has been idle for an extended period of time, its internal bearing surfaces are likely to become dry at points where the corrosion-preventive mixture has dried out or drained away from the bearings. Pre-oiling alleviates this problem.
REFERENCE: AC65-12A, Page 372

4086. ANSWER #4
Push-pull control rods are provided with an inspection hole for checking the proper thread engagement in the rod. When inspecting the control rods, a piece of safety wire should be inserted into the hole. If the wire will pass all the way through, there is not sufficient thread engagement.
REFERENCE: AC65-12A, Page 378

4087. ANSWER #4
The basic protection from detonation is provided in the design of the engine carburetor setting, which automatically supplies the rich mixtures required for detonation suppression at high power. Design factors, such as cylinder cooling, magneto timing, mixture distribution, supercharging, and carburetor setting are taken care of in the design and development of the engine and its method of installation in the aircraft.
REFERENCE: AC65-12A, Pages 444 and 445

4088. ANSWER #4
An unsupercharged recip engine operating at altitude will be less powerful than it is at sea level, because of the reduced density of the air and therefore the less potent fuel/air charge.
REFERENCE: ITP POWERPLANT, Chapter 1, Pages 4 and 6
                        AC65-12A, Page 111

4089. ANSWER #4
An extremely lean mixture either will not burn at all or will burn so slowly that combustion is not complete at the end of the exhaust stroke. The flame lingers in the cylinder and then ignites the contents in the intake manifold or the induction system when the intake valve opens. This causes an explosion known as backfiring.
REFERENCE: AC65-12A, Page 445

4090. ANSWER #3
When an engine is being pre-oiled, a line is disconnected or an opening made in the nose of the engine to allow oil to flow out. Oil flowing out of the engine indicates the completion of the pre-oiling operation, since the oil has now passed through the entire system.
REFERENCE: AC65-12A, Page 373

4091. ANSWER #4
When changing power settings, care should be taken to prevent damage to the engine by creating too high a manifold pressure for a given RPM. When power is to be increased, it is best to increase the RPM to the desired setting and then advance the throttle to the desired manifold pressure. When decreasing power, pull the throttle back until the manifold pressure is about one inch below the desired setting and then reduce the RPM with the propeller governor control.
REFERENCE: ITP POWERPLANT, Chapter 8, Page 67

4092. ANSWER #1
For a given RPM, the power output of the engine is less with the best-economy setting (auto-lean) than with the best-power mixture.
REFERENCE: AC65-12A, Pages 446 and 447 (Figure 10-34)

4093. ANSWER #2
When a fuel/air mixture does not contain enough fuel to consume all the oxygen, it is called a lean mixture. An extremely lean mixture either will not burn at all or will burn so slowly that combustion is not complete at the end of the exhaust stroke. The flame lingers in the cylinder and then ignites the contents in the intake manifold or the induction system when the intake valve opens. This causes an explosion known as backfiring.
REFERENCE: AC65-12A, Pages 113 and 445

4094. ANSWER #3
The critical point of detonation varies with the ratio of fuel to air in the mixture. Therefore, the detonation characteristic of the mixture can be controlled by varying the fuel/air ratio. At high power output, combustion pressures and temperatures are higher than they are at low or medium power. Therefore, at high power the fuel/air ratio is made richer than is needed for good combustion at medium or low power output. This is done because, in general, a rich mixture will not detonate as readily as a lean mixture.
REFERENCE: AC65-12A, Page 444

4095. ANSWER #4
If there is a small induction system air leak, it will be most noticeable at idle. The engine will probably tend to idle fast. At higher power settings, a small leak would probably not be noticable.
REFERENCE: ITP POWERPLANT, Chapter 1, Page 75
             AC65-12A, Page 9

4096. ANSWER #1
When changing power settings, care should be taken to prevent damage to the engine by creating too high a manifold pressure for a given RPM. When power is to be increased, it is best to increase the RPM to the desired setting and then advance the throttle to the desired manifold pressure. When decreasing power, pull the throttle back until the manifold pressure is about one inch below the desired setting and then reduce the RPM with the propeller governor control.
REFERENCE: ITP POWERPLANT, Chapter 8, Page 67

4097. ANSWER #4
Much of the condition within the cylinders of recip engines can be determined by studying the spark plugs. Normal operation is indicated by a spark plug having a relatively small amount of light brown or tan deposit on the nose of the center electrode insulator.
REFERENCE: ITP POWERPLANT, Chapter 1, Page 56

4098. ANSWER #1
In any piston engine, whether it is new or has been in service many hours, some amount of combustion chamber pressure gets by the piston rings and into the engine case. This pressure is vented to atmosphere through the crankcase breather. If the crankcase breather is plugged, the pressure will build up inside the crankcase.
REFERENCE: AC65-12A, Pages 286 and 287

4099. ANSWER #4
When there is too much valve clearance in a recip engine, the valves will not open as wide or remain open as long as they should. This reduces the overlap period.
REFERENCE: AC65-12A, Page 450

4100. ANSWER #2
Critical altitude means the maximum altitude at which, in standard atmosphere, it is possible to maintain, at a specified rotational speed, a specified power or a specified manifold pressure. The power referred to is brake horsepower.
REFERENCE: FAR 1

4101. ANSWER #4
The pressure in a gas turbine engine increases with each stage of the compressor. The pressure is the highest at the outlet of the compressor, which is known as the diffuser.
REFERENCE: ITP POWERPLANT, Chapter 2, Page 44
             AC65-12A, Page 66 and 67

4102. ANSWER #3

When the high energy gases leave the combustion section of a turbine engine, they enter the turbine section. The turbine section is made up of stationary and rotating airfoils. The stationary airfoils, or vanes, are known by a variety of names. These include turbine nozzle, turbine stator vanes, and nozzle diaphragm. The purpose of these vanes is to direct the high energy gases leaving the combustor into the rotating blades of the turbine. The nozzle diaphragm also increases the velocity of the gases.
REFERENCE: ITP POWERPLANT, Chapter 2, Page 48
                AC65-12A, Page 55

4103. ANSWER #3

A profile on a turbine engine compressor blade is a reduced thickness at the tip of the blade. If the tip of the blade happens to come in contact with the case, the profile is designed to wear away rather than the blade having a tendency to break.
REFERENCE: ITP POWERPLANT, Chapter 2, Page 39
                AC65-12A, Page 46

4104. ANSWER #2

On a dual axial or dual spool turbofan engine, the forward fan is bolted to the first compressor in the engine. The first compressor is known as the low-pressure compressor, or N1, because it imparts the least amount of pressure rise to the air.
REFERENCE: ITP POWERPLANT, Chapter 2, Pages 6 and 7
                AC65-12A, Page 104

4105. ANSWER #4

The abbreviation "P" stands for pressure, and the subscript t7 means total at station 7. Pt7 is the total pressure aft of the turbine.
REFERENCE: ITP POWERPLANT, Chapter 6, Pages 57 and 62

4106. ANSWER #2

When the high energy gases leave the combustion section of a turbine engine, they enter the turbine section. The turbine section is made up of stationary and rotating airfoils. The stationary airfoils, or vanes, are known by a variety of names. These include turbine nozzle, turbine stator vanes, and nozzle diaphragm. The purpose of these vanes is to direct the high energy gases leaving the combustor into the rotating blades of the turbine. The nozzle diaphragm also increases the velocity of the gases.
REFERENCE: ITP POWERPLANT, Chapter 2, Page 48
                AC65-12A, Page 55

4107. ANSWER #1

The combustion section of the turbine engine is where the fuel and air are mixed and then burned.
REFERENCE: ITP POWERPLANT, Chapter 2, Page 44
                AC65-12A, Page 48

4108. ANSWER #2

The combustion process in a turbine engine occurs at a constant pressure. It is known as the Brayton cycle. Although a pressure increase is normally associated with combustion, a pressure increase only happens for sure when combustion takes place in a closed space. The design of the turbine engine's combustor allows combustion to take place without an increase in pressure.
REFERENCE: ITP POWERPLANT, Chapter 2, Page 14
                AC65-12A, Page 67

4109. ANSWER #2

Increases in thrust relative to RPM are fairly proportional with a centrifugal compressor turbine engine. With an axial compressor turbine engine, however, the thrust increases more rapidly in the high RPM range than it does in the low RPM range.
REFERENCE: ITP POWERPLANT, Chapter 2, Page 24
                AC65-12A, Page 48

4110. ANSWER #2

A two-spool or dual spool turbine engine is one with two separately rotating units. The front compressor is called the low speed compressor, and the rear compressor is called the high speed. This type of engine has more operating flexibility than the single spool engine because the two compressors are free to find their own optimum RPM. At altitude, for example, the low speed compressor will increase in RPM because of the decrease in drag caused by the decrease in air density.
REFERENCE: ITP POWERPLANT, Chapter 2, Page 36

4111. ANSWER #3
When the high energy gases leave the combustion section of a turbine engine, they enter the turbine section. The turbine section is made up of stationary and rotating airfoils. The stationary airfoils, or vanes, are known by a variety of names. These include turbine nozzle, turbine stator vanes, and nozzle diaphragm. The purpose of these vanes is to direct the high energy gases leaving the combustor into the rotating blades of the turbine. The nozzle diaphragm also increases the velocity of the gases.
REFERENCE: ITP POWERPLANT, Chapter 2, Page 48
                         AC65-12A, Page 55

4112. ANSWER #3
The highest pressure in a gas turbine engine is at the outlet of the compressor, which is a section known as the diffuser. This same high pressure also exists at the entrance to the burner section.
REFERENCE: ITP POWERPLANT, Chapter 2, Page 44
                         AC65-12A, Pages 66 and 67

4113. ANSWER #1
Although the exhaust section of a turbine engine used in a subsonic airplane is convergent in shape, the exhaust cone attached to the rear frame of the turbine case makes the first part of the exhaust duct divergent in shape. In accordance with Bernoulli's theorm, this divergent shape causes the gas pressure to increase and the velocity to decrease.
REFERENCE: AC65-12A, Page 58

4114. ANSWER #2
The stator vane assembly at the discharge end of a typical axial-flow compressor is generally called exit guide vanes. The purpose of these vanes is to straighten the airflow and eliminate the turbulence before the air enters the combustor.
REFERENCE: ITP POWERPLANT, Chapter 2, Page 41
                         AC65-12A, Page 46

4115. ANSWER #4
The purpose of the turbine section in a gas turbine engine is to extract energy from the gases leaving the combustor and convert that energy into rotary motion. The energy extracted is used to drive the compressors, the fan, the prop, the rotor blades, etc., according to the type of turbine engine being looked at.
REFERENCE: ITP POWERPLANT, Chapter 2, Page 47
                         AC65-12A, Page 53

4116. ANSWER #1
One of the critical factors to observe when starting a turbine engine is the exhaust gas temperature. If the exhaust gas temperature exceeds the specified limits during an attempted start, something called a "hot start" has occurred.
REFERENCE: ITP GENERAL, Chapter 7, Page 26
                         AC65-9A, Page 494

4117. ANSWER #3
In a dual spool axial compressor turbine engine the second compressor, or N2, is driven by the first turbine. The second turbine shafts through the first turbine and its compressor to drive the first compressor.
REFERENCE: ITP POWERPLANT, Chapter 2, Page 35

4118. ANSWER #1
When marking damaged areas in the hot section of a turbine engine, a lead pencil should never be used. The carbon alloy in the lead pencil could cause an intergrannular attack in the metal, which could result in a reduction of the material's strength.
REFERENCE: ITP POWERPLANT, Chapter 2, Page 66
                         AC65-12A, Page 477

4119. ANSWER #2
In many turbofan engines, the fan is bolted to the low speed compressor (N1), and is therefore driven by the same turbine. Some fan engines drive the fan with a separate turbine, as identified in answer #3, but this is not as common.
REFERENCE: ITP POWERPLANT, Chapter 2, Pages 6 and 7
                         AC65-12A, Page 104

4120. ANSWER #2
When a turbine engine fails to reach idle RPM during an attempted start, the engine is said to have experienced a hung start. This is generally caused by shutting off the starter too soon, or by improper fuel metering.
REFERENCE: ITP GENERAL, Chapter 7, Page 26
              AC65-9A, Page 494

4121. ANSWER #2
The two main sections of a turbine engine, for inspection purposes, are the hot section and the cold section. The cold section extends back to the diffuser. The hot section includes the combustor and the turbine.
REFERENCE: ITP POWERPLANT, Chapter 2, Page 58
              AC65-12A, Page 472

4122. ANSWER #4
The two basic elements of the turbine section of a turbine engine are the stator and the rotor. The stator is the stationary airfoils, located in front of the rotor. The stator is also known as the turbine nozzle or nozzle diaphragm. The rotor is the rotating airfoils, or turbine blades.
REFERENCE: ITP POWERPLANT, Chapter 2, Pages 48 and 49
             AC65-12A, Page 53

4123. ANSWER #3
The exhaust cone of a turbine engine attaches to the rear of the turbine case and extends into the exhaust duct. Because of its shape, it causes the front of the exhaust duct to be a divergent duct. This diverging shape causes the velocity of the gases to be decreased and the pressure to be increased. By slowing the velocity of the gases down for an instant, the swirling motion they have coming off the turbine is reduced before they exit out the exhaust duct.
REFERENCE: AC65-12A, Page 58

4124. ANSWER #4
The two parts that make up a centrifugal compressor are the impeller and the diffuser.
REFERENCE: ITP POWERPLANT, Chapter 2, Page 32
             AC65-12A, Page 43

4125. ANSWER #3
After the fuel control has been replaced on a turbine engine, it is often necessary to retrim the engine. This is not required or necessary, however, on all turbine engines. Some present day turbine engines, such as a G.E. T700, can have the fuel control replaced without trimming being necessary.
REFERENCE: ITP POWERPLANT, Chapter 6, Page 62

4126. ANSWER #1
The most satisfactory and popular design of turbine blade bases is the fir-tree design. The fir-tree, when it mates with the corresponding fir-tree shape in the turbine disc, provides the strongest means of attachment.
REFERENCE: ITP POWERPLANT, Chapter 2, Page 49
             AC65-12A, Page 56

4127. ANSWER #2
A double-sided centrifugal compressor has vanes on both sides of the impeller.
REFERENCE: ITP POWERPLANT, Chapter 2, Page 33
             AC65-12A, Pages 43 to 45

4128. ANSWER #4
In the starting sequence of a turbine engine, the engine is rotated with the starter to a certain percent RPM before the ignition and then the fuel are supplied to the engine. The first indication in the cockpit that a successful start has occurred is the rise in temperature indicated on the exhaust gas temperature gage.
REFERENCE: AC65-12A, Page 487

4129. ANSWER #3
One of the advantages of a dual spool axial compressor over a single spool is the ability of the two separate compressors to seek their own optimum RPM. By allowing the two compressors to rotate at different speeds, the compression ratio and therefore efficiency of the engine can be increased.
REFERENCE: ITP POWERPLANT, Chapter 2, Page 36

4130. ANSWER #3
The two basic types of turbine blades in use today are the reaction and the impulse. In actuality, turbine engines today use blades which are combination impulse-reaction.
REFERENCE: ITP POWERPLANT, Chapter 2, Page 48

4131. ANSWER #3
For a typical turboprop engine in service today, the propeller accounts for approximately 75 to 85 percent of the engine's thrust output. The other 15 to 25 percent comes from the exhaust gases.
REFERENCE: ITP POWERPLANT, Chapter 2, Page 10

4132. ANSWER #3
Although an axial flow compressor in a turbine engine does not give as high a compression rise per stage as a centrifugal compressor, its ability to use multiple stages and take advantage of ram effect give it high peak efficiencies.
REFERENCE: ITP POWERPLANT, Chapter 2, Page 36
             AC65-12A, Page 48

4133. ANSWER #3
The stationary airfoils in the axial flow compressor are most appropriately called stator vanes. Their purpose is twofold: (1) to direct the airflow from stage to stage and (2) to increase the pressure of the air.
REFERENCE: ITP POWERPLANT, Chapter 2, Page 38
             AC65-12A, Page 46

4134. ANSWER #1
In a turbine engine, the diffuser section is at the outlet of the compressor. The purpose of the diffuser is to reduce the velocity of the gases and to increase their pressure.
REFERENCE: ITP POWERPLANT, Chapter 2, Page 44
             AC65-12A, Page 61

4135. ANSWER #4
Stress rupture cracks on turbine blades usually appear as minute hairline cracks on or across the leading or trailing edge at a right angle to the edge length. Stress rupture cracks generally develop because of a stress being present in the blade and, due to the elevated temperatures and centrifugal loading, they relieve in the form of a crack.
REFERENCE: ITP POWERPLANT, Chapter 2, Page 63
             AC65-12A, Page 479

4136. ANSWER #1
A can-type combustion chamber, because it is located externally around the circumference of the engine, could be removed and installed as one unit during routine maintenance on the engine.
REFERENCE: ITP POWERPLANT, Chapter 2, Page 46
             AC65-12A, Page 49

4137. ANSWER #4
The diffuser section of a turbine engine is located at the outlet of the compressor and the inlet of the combustor.
REFERENCE: ITP POWERPLANT, Chapter 2, Pages 17 and 44

4138. ANSWER #3
The mechanical blockage and the aerodynamic blockage are the most common types of thrust reversers used on turbine-powered aircraft. The mechanical blockage type usually deploys aft of the jet nozzle, while the aerodynamic blockage type deploys inside the tailpipe.
REFERENCE: ITP POWERPLANT, Chapter 7, Page 40
             AC65-12A, Page 103

4139. ANSWER #3
Stress rupture cracks on turbine blades usually appear as minute hairline cracks on or across the leading or trailing edge at a right angle to the edge length. Stress rupture cracks generally develop because of a stress being present in the blade and, due to the elevated temperatures and centrifugal loading, they relieve in the form of a crack.
REFERENCE: ITP POWERPLANT, Chapter 2, Page 63
             AC65-12A, Page 479

4140. ANSWER #2

When inspecting a turbine engine, it is normal to find more distress in the turbine section than in the compressor section. Although both are subject to high centrifugal loading, the turbine section operates at much higher temperatures.
REFERENCE: ITP POWERPLANT, Chapter 2, Page 63
                        AC65-12A, Page 479

4141. ANSWER #3

The greatest stress which a turbine engine is subjected, is the high internal operating temperatures. The ultimate limit on how much thrust a given engine can produce is determined by how high the turbine inlet temperature can be.
REFERENCE: ITP POWERPLANT, Chapter 5, Page 33
                        EA-TEP, Page 119
                        AC65-12A, Page 486

4142. ANSWER #4

The turbine shaft is usually joined to the compressor rotor of a centrifugal compressor turbine engine by a splined coupling.
REFERENCE: EA-TEP, Page 39
                        AC65-12A, Page 56

4143. ANSWER #4

During the operation of a turbine engine, the turbine inlet temperature is the most critical variable. Although the temperature gage in the cockpit might be indicating exhaust gas temperature, this is only an indication of what the inlet temperature must be. An overtemperature of only a few degrees might require an engine to be sent to overhaul, so the temperature must be monitored very closely.
REFERENCE: ITP POWERPLANT, Chapter 5, Page 33
                        AC65-12A, Page 486

4144. ANSWER #3

The use of shrouded turbine rotor blades has done much to reduce blade vibration and to improve the efficiency of the turbine. Because the shrouds on the tips of the blades are in contact with each other, each blade tends to support the other, thereby reducing vibration. The shrouds also help to control the airflow through the turbine, not allowing air to escape over the tip of the blade. This helps improve the efficiency of the turbine.
REFERENCE: ITP POWERPLANT, Chapter 2, Page 49
                        AC65-12A, Page 57

4145. ANSWER #3

Of the choices given in this question, the split-spool, axial flow compressor has the most to offer. It is easier to start an engine with this type of compressor because the starter only rotates the N2 compressor. The high altitude performance is better because the two separately rotating compressors are able to seek their own optimum RPM. At altitude where the air is less dense, the N1 compressor is able to speed up and thereby increase the compression of the engine.
REFERENCE: ITP POWERPLANT, Chapter 2, Page 36
                        AC65-12A, Page 48

4146. ANSWER #4

In order to maintain the balance of the turbine assembly, when a turbine blade is removed for inspection, it must be re-installed in the same position it came from.
REFERENCE: ITP POWERPLANT, Chapter 2, Page 63
                        AC65-12A, Page 479

4147. ANSWER #2

Although the peak efficiency of the centrifugal compressor is not as great as the axial-flow type, it does give a higher pressure rise per stage. Modern day centrifugal compressors are giving as much as 8 or 10 to 1 compression per stage, while axial-flow compressors give approximately 1.3 to 1 per stage.
REFERENCE: ITP POWERPLANT, Chapter 2, Page 34
                        AC65-12A, Page 48

4148. ANSWER #3

The highest temperature in a gas turbine engine occurs in the combustor in the middle of the flame zone. At this point in the engine, however, the high temperature is shielded from the metal by a cooling blanket of air. At the inlet to the turbine the high combustor temperatures have been cooled considerably, but at this point there is direct heat to metal contact.
REFERENCE: EA-TEP, Page 38

4149. ANSWER #1
The axial-flow compressor assembly is made up of two elements, the rotor and the stator. The rotor is the rotating airfoils and the stator is the stationary airfoils.
REFERENCE: ITP POWERPLANT, Chapter 2, Page 38
AC65-12A, Page 45

4150. ANSWER #2
The two types of centrifugal compressor impellers are the single entry and the double entry. The single entry has vanes on only one side of the impeller, while the double entry has vanes on both sides of the impeller.
REFERENCE: ITP POWERPLANT, Chapter 2, Page 33
AC65-12A, Page 43

4151. ANSWER #4
Between each row of rotating blades in an axial-flow compressor is a set of stationary airfoils called stator vanes. The stator vanes direct the air between stages and they act as diffusers to increase the pressure of the air.
REFERENCE: ITP POWERPLANT, Chapter 2, Page 38
AC65-12A, Page 45

4152. ANSWER #4
Standard day sea level pressure is 29.92 inches of mercury or 14.7 pounds per square inch.
REFERENCE: ITP GENERAL, Chapter 1, Page 73
AC65-9A, Page 241

4153. ANSWER #4
Standard day sea level temperature is 15 degrees Centigrade or 59 degrees Fahrenheit.
REFERENCE: AC65-9A, Page 241

4154. ANSWER #4
When turbine blades are subjected to excessive temperatures, stress rupture cracks are likely to develop. Stress rupture cracks usually appear as minute hairline cracks on or across the leading or trailing edge at a right angle to the edge length.
REFERENCE: ITP POWERPLANT, Chapter 2, Page 63
AC65-12A, Page 479

4155. ANSWER #2
Of the answers given, number two appears to be the best. Answer number four would be correct if the word "decrease" were to read "increase".
REFERENCE: ITP POWERPLANT, Chapter 2, Page 38
AC65-12A, Page 46

4156. ANSWER #4
There are two ways of viewing this question. We know for sure that dirty compressor blades are not as efficient as clean ones, so they will not be able to pump as much air for a given RPM. Because the compressor is not pumping as much air, for a given RPM the exhaust gas temperature is going to be high. In terms of exhaust gas temperature (EGT), for a given EGT the RPM is going to be low. Although there could be two answers, depending on how you look at the question, answer #4 is the best choice.
REFERENCE: ITP POWERPLANT, Chapter 2, Page 56
AC65-12A, Page 472

4157. ANSWER #4
The two types of compressors used most often in gas turbine engines are the axial-flow and the centrifugal flow. In some engines, a combination axial-centrifugal compressor is used.
REFERENCE: ITP POWERPLANT, Chapter 2, Pages 32 and 35
AC65-12A, Page 43

4158. ANSWER #1
The use of shrouded turbine rotor blades has done much to reduce blade vibration and to improve the efficiency of the turbine. Because the shrouds on the tips of the blades are in contact with each other, each blade tends to support the other, thereby reducing vibration. The shrouds also help to control the airflow through the turbine, not allowing air to escape over the tip of the blade. This helps improve the efficiency of the turbine.
REFERENCE: ITP POWERPLANT, Chapter 2, Page 49
AC65-12A, Page 57

4159. ANSWER #1

In a dual axial-flow compressor turbine engine, the first stages of turbine are used to drive the second compressor (N2). The second stages of turbine shaft through the first and drive the first compressor (N1).
REFERENCE: ITP POWERPLANT, Chapter 2, Page 35

4160. ANSWER #1

If a turbine engine catches fire during a starting attempt, the fuel should be shut off and the engine rotated with the starter. By continuing to rotate the engine, the fire should be drawn through the engine and discharged out the tailpipe.
REFERENCE: ITP POWERPLANT, Chapter 2, Pages 79 and 80
                      AC65-12A, Page 488

4161. ANSWER #3

A typical sequence for starting a turbine engine is as follows:
1.   Engage the starter.
2.   At 15% RPM, turn on the ignition.
3.   After the ignition has been turned on, turn on the fuel.
4.   Disengage the starter after the engine has reached a self-accelerating speed.
REFERENCE: ITP POWERPLANT, Chapter 2, Page 79
                      AC65-12A, Page 486

4162. ANSWER #2

Inflight turbine engine flame-outs are usually caused by interruption of the inlet airflow. As the airflow into the inlet is decreased, the pressure build-up in the engine diffuser decreases. Combustion chamber pressures now become greater than the pressure in the diffuser and the flame front backs up into the compressor. A massive compressor stall is the result and very likely an engine flame-out.
REFERENCE: ITP POWERPLANT, Chapter 2, Page 45

4163. ANSWER #1

In a gas turbine engine, bleed air valves help keep the operation of the compressor stable during low thrust conditions. To make axial-flow compressors as efficient as possible, they are operated with a blade angle of attack which is close to a stall condition. To prevent the compressor from loading up and trying to pump too much air when it is at low RPM, bleed valves are installed which dump some of the compressor air overboard. This allows the compressor to accelerate more rapidly and alleviates a potential stall problem. As the engine accelerates to a higher RPM, the bleed air valves close and allow all the air to flow through the engine.
REFERENCE: ITP POWERPLANT, Chapter 7, Page 20

4164. ANSWER #3

Most modern day gas turbine engines use a dual-spool type of compressor, utilizing two axial-flow rotors or one axial and one centrifugal. A big advantage of the dual-spool compressor is the ability of the first compressor (N1) to seek its own best operating speed. When the engine is operated at altitude where the air is less dense, the reduced drag on the N1 compressor allows it to speed up and add to the compression of the engine.
REFERENCE: ITP POWERPLANT, Chapter 2, Page 36

4165. ANSWER #1

The purpose of an inlet guide vane assembly in a gas turbine engine is to direct the airflow into the first stage rotor blades at the proper angle and to impart a swirling motion to the air in the direction of rotation. Inlet guide vanes really apply to an axial-flow compressor and not a centrifugal compressor as this question asks about.
REFERENCE: ITP POWERPLANT, Chapter 2, Pages 39 and 40
                      AC65-12A, Page 46

4166. ANSWER #1

When inspecting the hot section of a turbine engine, the exhaust cone and tailpipe should be inspected for cracks, warping, buckling, or hotspots. Hotspots on the tail cone are a good indication of a malfunctioning fuel nozzle or combustion chamber. If a fuel nozzle is spraying a solid stream of fuel instead of an atomized spray, the fuel will still be burning when it contacts the tail cone and it will produce a burn mark. A combustion chamber which is not properly controlling the flame zone might allow the flame to come in contact with the tail cone.
REFERENCE: AC65-12A, Page 482

4167. ANSWER #1
Stator vanes in an axial-flow compressor separate each rotating set of blades. Their purpose is to direct the airflow at the proper angle to the next set of rotating blades and, because they act as diffusers, they also increase the pressure of the air.
REFERENCE: ITP POWERPLANT, Chapter 2, Page 38
             AC65-12A, Page 46

4168. ANSWER #4
According to Bernoulli's theorem, when subsonic airflow at a constant flow rate passes through a convergent nozzle, its velocity increases and its pressure decreases. This happens because when one unit enters the nozzle, one unit has to leave the nozzle. Because a convergent nozzle is getting smaller from front to back, the unit in the back is being squeezed down and therefore has to speed up to get out to make room to the next unit coming in.
REFERENCE: ITP POWERPLANT, Chapter 2, Page 50
             AC65-12A, Page 67

4169. ANSWER #4
According to Bernoulli's theorem, when subsonic airflow at a constant flow rate passes through a divergent nozzle, its velocity decreases and its pressure increases. This happens because when one unit enters the nozzle, one unit has to leave the nozzle. Because a divergent nozzle is getting larger from front to back, the unit in the back is spreading out and therefore has to slow down to exit as a unit as the next unit comes in.
REFERENCE: ITP POWERPLANT, Chapter 2, Pages 12 and 13
             AC65-12A, Page 67

4170. ANSWER #2
As identified in the explanation to question #4168, the velocity of air increases when it passes through a convergent nozzle. Because the total energy in the air does not change, if the velocity energy goes up, the pressure energy must go down a proportional amount.
REFERENCE: ITP POWERPLANT, Chapter 2, Pages 12 and 13
             AC65-12A, Page 67

4171. ANSWER #3
As identified in the explanation to question #4169, the velocity of air decreases when it passes through a divergent nozzle. Because the total energy in the air does not change, if the velocity energy goes down, the pressure energy must go up a proportional amount.
REFERENCE: ITP POWERPLANT, Chapter 2, Pages 12 and 13
             AC65-12A, Page 67

4172. ANSWER #3
Anti-icing of turbine engine air inlets is normally done by using air bled off the compressor of the engine.
REFERENCE: ITP POWERPLANT, Chapter 7, Page 23

4173. ANSWER #2
A typical sequence for starting a turbine engine is as follows:
1.   Engage the starter.
2.   At 15% RPM, turn on the ignition.
3.   After the ignition has been turned on, turn on the fuel.
4.   Disengage the starter after the engine has reached a self-accelerating speed. The speed is close to the idle RPM.
REFERENCE: ITP POWERPLANT, Chapter 2, Page 79
             AC65-12A, Page 486

4174. ANSWER #4
Although an axial-flow compressor does not give as much pressure ratio per stage as the centrifugal compressor, it is capable of a greater total pressure rise through the use of multiple stages.
REFERENCE: ITP POWERPLANT, Chapter 2, Page 36
             AC65-12A, Page 48

4175. ANSWER #1
Included in some inlet ducts for turbine engines with centrifugal compressors, as a necessary part of the plenum chamber, are auxiliary air-intake doors (blow-in doors). These blow-in doors admit air to the engine compartment during ground operation, when air requirements for the engine are in excess of the airflow through the inlet ducts.
REFERENCE: AC65-12A, Page 43

4176. ANSWER #4
A double entry centrifugal compressor is one that has vanes on both sides of the impeller.
REFERENCE: ITP POWERPLANT, Chapter 2, Page 33
                  AC65-12A, Page 43

4177. ANSWER #3
The purpose of the turbine section in a gas turbine engine is to extract energy from the gases coming off the combustor. This energy is converted into rotary motion of the turbine to drive the compressor, fan, prop, rotor shaft, and gearbox, according to the type of engine in use.
REFERENCE: ITP POWERPLANT, Chapter 2, Page 47
                  AC65-12A, Page 53

4178. ANSWER #3
Stator vanes in an axial-flow compressor separate each rotating set of blades. Their purpose is to direct the airflow at the proper angle to the next set of rotating blades and, because they act as diffusers, they also increase the pressure of the air.
REFERENCE: ITP POWERPLANT, Chapter 2, Page 38
                  AC65-12A, Page 46

4179. ANSWER #4
The three main sections of a gas turbine engine are the compressor, combustion, and the turbine.
REFERENCE: ITP POWERPLANT, Chapter 2, Page 5
                  AC65-12A, Page 39

4180. ANSWER #4
The two singular types of turbine blades used in aircraft gas turbine engines are the impulse and the reaction. Most engines, however, use a combination blade known as an impulse-reaction turbine blade. With this blade, the lower part is an impulse shape and the upper part is a reaction shape.
REFERENCE: ITP POWERPLANT, Chapter 2, Pages 48 and 49

4181. ANSWER #1
The compression ratio of an engine using an axial-flow compressor varies according to:
1.   The number of stages
2.   The efficiency of the compressor
3.   The pressure rise per stage
REFERENCE: ITP POWERPLANT, Chapter 2, Page 37
                  AC65-12A, Page 45

4182. ANSWER #3
The three main sections of a gas turbine engine are the compressor, combustion, and the turbine.
REFERENCE: ITP POWERPLANT, Chapter 2, Page 5
                  AC65-12A, Page 39

4183. ANSWER #2
If the engine power parameters are not changing, but the engine oil temperature is high, there is a good possibility that a bearing is in the process of failing. As the bearing gets closer to failing, the power parameters of the engine will change. In the early stages of distress, however, the only sign may be an increase in oil temperature.
REFERENCE: ITP POWERPLANT, Chapter 8, Page 32
                  AC65-12A, Page 491

4184. ANSWER #3
Regardless of the type, all fuel controls accomplish essentially the same function, but some sense more engine variables than others. Some of the variables which may be sensed by the fuel control are power lever position, engine RPM, either compressor inlet pressure or temperature, and burner pressure or compressor discharge pressure. There is no mixture control on a turbine engine.
REFERENCE: ITP POWERPLANT, Chapter 6, Page 51
                  AC65-12A, Page 149

4185. ANSWER #3
Newton's First Law of Motion states that every body persists in its state of rest, or of motion in a straight line, unless acted upon by some outside force.
REFERENCE: ITP GENERAL, Chapter 1, Page 66
                  ITP POWERPLANT, Chapter 2, Page 11
                  AC65-9A, Page 254

4186. ANSWER #3
Because of the extremely high temperatures present, the hot section of a turbine engine is especially susceptible to cracking.
REFERENCE: ITP POWERPLANT, Chapter 2, Page 62
             AC65-12A, Pages 477 to 479

4187. ANSWER #1
The atmosphere near the ground is filled with tiny particles of dirt, oil, soot, and other foreign matter. A large volume of air is introduced into the compressor, and centrifugal force throws the dirt particles outward so that they build up to form a coating on the casing, the vanes, and the compressor blades. The turbine blades are not susceptible to this problem because of the high temperatures present in the engine hot section.
REFERENCE: AC65-12A, Page 472

4188. ANSWER #4
Galling is a transfer of metal from one surface to another, usually caused by severe rubbing.
REFERENCE: AC65-12A, Page 476

4189. ANSWER #2
Regardless of the type, all fuel controls accomplish essentially the same function, but some sense more engine variables than others. Some of the variables which may be sensed by the fuel control are power lever position, engine RPM, either compressor inlet pressure or temperature, and burner pressure or compressor discharge pressure. There is no mixture control on a turbine engine.
REFERENCE: ITP POWERPLANT, Chapter 6, Page 51
             AC65-12A, Page 149

4190. ANSWER #3
Although the AC65-12A would indicate that the engine being out of trim is the answer to this question, it is not a correct answer. Trimming the engine means adjusting the idle RPM and the maximum RPM, or the takeoff EPR if it is an EPR rated engine. In this question, the desired EPR for takeoff has been reached. The fact that the exhaust gas temperature (EGT) is too high has nothing to do with trimming. The most likely cause of high EGT when the takeoff EPR has been reached is a dirty or damaged compressor. This condition decreases the airflow through the engine, making it impossible to get takeoff EPR without advancing the power lever more and adding more fuel. The end result is a high EGT.
REFERENCE: AC65-12A, Page 472

4191. ANSWER #1
The Brayton cycle is what describes the combustion process in a turbine engine. It is known as the constant pressure cycle because the pressure across combustion in a turbine engine remains relatively constant.
REFERENCE: ITP POWERPLANT, Chapter 2, Page 14
             AC65-12A, Page 66

4192. ANSWER #1
When a turbine engine uses water injection, the fluid is injected at the compressor inlet or diffuser, or both.
REFERENCE: ITP POWERPLANT, Chapter 6, Page 72
             AC65-12A, Page 167

4193. ANSWER #2
Continued and/or excessive heat and centrifugal force acting on turbine engine compressor blades may cause them to stretch or grow. The problem of blade growth in a turbine engine, however, is really in the turbine section where the heat is extreme.
REFERENCE: ITP POWERPLANT, Chapter 2, Page 64
             AC65-12A, Page 476

4194. ANSWER #1
The angle of attack on the compressor blades in a turbine engine is affected by three things, the RPM of the compressor, the direction of the airflow coming off the stator vanes, and the velocity of the airflow coming off the stator vanes.
REFERENCE: ITP POWERPLANT, Chapter 2, Page 42

4195. ANSWER #1
The compression ratio of an engine using an axial-flow compressor varies according to:
1.   The number of stages
2.   The efficiency of the compressor
3.   The pressure rise per stage
REFERENCE: ITP POWERPLANT, Chapter 2, Page 37
               AC65-12A, Page 45

4196. ANSWER #2
The power produced by a turbine engine is proportional to the stagnation density at the inlet. The factors which affect the density at the inlet are altitude, airspeed, and outside air temperature.
REFERENCE: ITP POWERPLANT, Chapter 2, Page 23
               AC65-12A, Page 68

4197. ANSWER #3
The three most important factors affecting the thermal efficiency of a turbine engine are turbine inlet temperature, compression ratio, and the efficiency of the compressor and turbine. Other factors that affect thermal efficiency are compressor inlet temperature and burner efficiency.
REFERENCE: ITP POWERPLANT, Chapter 2, Page 22
               AC65-12A, Page 67

4198. ANSWER #4
The number of turbine wheels used in a gas turbine engine is determined by the amount of energy which must be extracted to drive the compressor, possibly a fan or prop, and the gearboxes. A turbofan or turboprop engine will have more stages of turbine than does a turbojet, because additional energy is needed to drive the fan or prop.
REFERENCE: ITP POWERPLANT, Chapter 2, Page 48
               AC65-12A, Page 57

4199. ANSWER #1
The exhaust section of a turbine engine installed in a subsonic aircraft will have a converging shape. The converging shape increases the velocity of the exhaust gases. If the engine were installed in a supersonic aircraft, the exhaust section would have a converging-diverging (C-D) shape.
REFERENCE: ITP POWERPLANT, Chapter 2, Page 50
               AC65-12A, Page 58

4200. ANSWER #4
The type of combustion chambers used in gas turbine engines are the can, can-annular, and annular. In modern day engines, the annular is the most popular.
REFERENCE: ITP POWERPLANT, Chapter 2, Page 46
               AC65-12A, Page 49

4201. ANSWER #3
A cool-off period is often needed for a gas turbine engine before shutting it down to allow the turbine wheel to cool before the case contracts around it. The rate at which the case will cool compared to the turbine wheel and blades is much different. Not only are they made of different materials, but they are subject to a different cooling environment.
REFERENCE: AC65-12A, Page 489

4202. ANSWER #2
Regardless of the number of combustion chambers an engine uses, 2 igniter plugs are normally used. They are usually located at the four and eight o'clock position in an annular combustor and in the cans at approximately four and eight o'clock with can or can-annular combustors.
REFERENCE:ITP POWERPLANT, Chapter 3, Page 44
               AC65-12A, Page 230

4203. ANSWER #1
Most turbines are open at the outer perimeter of the blades; however, a second type called the shrouded turbine is sometimes used. The shrouded turbine blades, in effect, form a band around the outer perimeter of the turbine wheel. This improves efficiency and vibration characteristics, and permits lighter stage weights.
REFERENCE: ITP POWERPLANT, Chapter 2, Page 49
               AC65-12A, Page 57

4204. ANSWER #4
Creep is a term used to describe the permanent elongation which occurs to rotating parts. Creep is most pronounced in turbine blades because of the heat loads and centrifugal loads imposed during operation.
REFERENCE: ITP POWERPLANT, Chapter 2, Page 64

4205. ANSWER #2

The pressurization and dump valve is part of the fuel system in an engine which uses a dual line duplex fuel nozzle. The pressurization part of the valve controls the separation of fuel between the primary and secondary orifices in the fuel nozzle. It allows only primary fuel to flow during engine starting and low power operation, but adds secondary fuel flow when the engine reaches higher power operations.

REFERENCE: ITP POWERPLANT, Chapter 6, Pages 79 and 80
                AC65-12A, Page 173

4206. ANSWER #3

The highest pressure in a gas turbine engine is at the compressor outlet. This point in the engine is known as the diffuser.

REFERENCE: ITP POWERPLANT, Chapter 2, Page 44
                AC65-12A, Pages 66 and 67

4207. ANSWER #4

The jet nozzle of a gas turbine engine is the opening at the end of the tailpipe through which the gases escape to atmosphere.

REFERENCE: EA-TEP, Page 41
                AC65-12A, Page 58

4208. ANSWER #1

The model of engine identified in this question is 0-690, serial number 5863-40, with 283 hours' time in service. Paragraph "B" in the AD applies to this engine because it identifies model 0-690 engines with serial numbers 5265-40 to 6129-40. Paragraph "1" also applies because it identifies engines with more than 275 hours' time in service. Paragraph "C" also applies to this engine, but it was not a possible choice along with "B" and "1".

REFERENCE: AC39-7A

4209. ANSWER #3

The Aircraft Specifications or Type Certificate Data Sheet is the document which lists the approved engines and propellers for an aircraft. If there is more than one approved propeller for the Cessna 180, it will be listed in one of these documents.

REFERENCE: ITP GENERAL, Chapter 7, Page 31
                AC65-9A, Page 465

4210. ANSWER #2

The exhaust gas temperature (EGT) of a turbine engine is an operating limit and used to monitor the mechanical integrity of the turbines, as well as to check engine operating conditions. If there is damage to the turbine section of an engine, it will show up as an increased EGT because a damaged turbine will not pick off as much heat.

REFERENCE: EA-TEP, Page 119
                AC65-12A, Page 486

4211. ANSWER #1

Many piston engine powered aircraft use the engine exhaust heat as a source of cabin heat. They do so by using a jacket around the exhaust system so the air passing through the jacket can be heated by the hot exhaust manifold. When this type of system is used, it must be inspected on a regular basis to ensure that no leaks exist in the exhaust system. Even a small leak would introduce carbon monoxide into the cabin and cause carbon monoxide poisoning.

REFERENCE: ITP AIRFRAME, Chapter 7, Page 24

4212. ANSWER #4

The two statements found in this question come word for word from the AC65-9A. Statement #1 is definitely true because AD's are Federal Aviation Regulations and they must be complied with unless specific exemption is granted. Statement #2 is also true on most occasions, because emergency AD's most often do require immediate compliance. Some emergency AD's have been issued, however, which did not require immediate compliance. There is a problem with statement #2, but for the sake of this question, it should be answered as true.

REFERENCE: AC65-9A, Page 465
                AC39-7A

4213. ANSWER #3

FAR 43 Appendix "D" contains the minimum checklist for a 100-hour inspection of an engine.

REFERENCE: FAR 43, Appendix "D"

4214. ANSWER #1

After an AD has become effective, it must be complied with as specified in the AD.

REFERENCE: AC39-7A

4215. ANSWER #3
To find out what engines a particular propeller is adaptable to, a mechanic would need to look in the propeller's Type Certificate Data Sheet.
  REFERENCE: ITP GENERAL, Chapter 7, Page 38

4216. ANSWER #2
On a 100-hour inspection of an engine, Appendix "D" of FAR 43 requires that a cylinder compression check be performed. This is done to check the internal integrity of the engine's cylinders.
REFERENCE: FAR 43, Appendix "D"

4217. ANSWER #1
In the designation for an engine R985-22, the "985" indicates the total piston displacement of the engine in cubic inches.
REFERENCE: ITP POWERPLANT, Chapter 1, Pages 26 and 27

4218. ANSWER #1
Cylinder barrel chrome plating processes are approved by the FAA under the Supplemental Type Certificate method. Agencies whose processes are approved are listed in the Summary of Supplemental Type Certificates.
REFERENCE: AC43.13-1A, Pages 271 and 272

4219. ANSWER #3
"CAA Spec 5E4" indicates a Civil Aeronautics Authority specification. The meaning of this specification could be found in the Aircraft Engine Specifications.
REFERENCE: ITP GENERAL, Chapter 7, Pages 34 to 36

4220. ANSWER #2
A bent nitrided crankshaft should not be straightened. Any attempt to do so will result in rupture of the nitrided surface of the bearing journals, a condition that will cause eventual failure of the crankshaft.
REFERENCE: AC65-12A, Page 428

4221. ANSWER #1
Flaking is defined as the breaking loose of small pieces of metal or coated surfaces. It is usually caused by defective plating or excessive loading.
REFERENCE: AC65-12A, Page 412

4222. ANSWER #2
Each airplane which is type certificated must be equipped with an engine which is type certificated and, if it is prop driven, a type certificated propeller.
REFERENCE: FAR 23.903 (a)

4223. ANSWER #4
Galling is defined as a severe condition of chafing or fretting in which a transfer of metal from one part to another occurs. It is usually caused by a slight movement of mated parts having limited relative motion and under high loads.
REFERENCE: AC65-12A, Pages 412 and 413

4224. ANSWER #2
Brinelling is defined as one or more indentations on bearing races usually caused by high loads or application of force during installation or removal. The indentations are rounded or spherical due to the impression left by the contacting balls or rollers of the bearing.
REFERENCE: AC65-12A, Page 412

4225. ANSWER #3
When an engine is being inspected, it is important to make sure that it conforms to its original type design. The original type design is identified in the Aircraft Engine Specifications or the Type Certificate Data Sheet. Since 1959, the document has been called the Type Certificate Data Sheet. The Aircraft Engine Specifications would be used for engines certified prior to 1959.
REFERENCE: AC65-9A, Page 465

4226. ANSWER #3
Although a certified A&P mechanic is authorized to perform a major repair, it takes a person with the inspection authorization to return the repair to service.
REFERENCE: FAR 65.95 (a) (1)

4227. ANSWER #2

A list of what would constitute a powerplant major repair is provided in FAR 43, Appendix "A". According to FAR 1, anything that is not a major repair is a minor repair. Appendix "A" of FAR 43 is only a guide, it does not identify all the possibilities. It does not, for example, give any information about turbine engine repair in regards to major or minor.
REFERENCE: FAR 43, Appendix "A"

4228. ANSWER #2

FAR 23 prescribes the airworthiness standards for the issue of type certificates, and changes to those certificates, for small airplanes in the normal, utility, and acrobatic categories that have a passenger seating configuration, excluding pilot seats, of nine seats or less.
REFERENCE: FAR 23.1 (a)

4229. ANSWER #4

Manufacturer's manuals, such as overhaul and maintenance manuals, are acceptable to use when performing a major repair to an engine providing they are FAA approved. FAA approval must be stamped on the manual before it can be used as approved data.
REFERENCE: AC65-19B

4230. ANSWER #3

When a major repair is performed on an engine, an entry must be made in the engine's maintenance records and an FAA Form 337 must be filled out. One copy of the Form 337 stays with the maintenance records and a second copy is sent to the FAA. An exception to this would be if the repair was done by a certified repair station. If approved data was used, the repair station would not be required to fill out a Form 337.
REFERENCE: FAR 43.9; FAR 43, Appendix "B"

4231. ANSWER #2

When an engine has been subjected to sudden stoppage and a crankshaft run-out check is deemed necessary, it should be done in accordance with the manufacturer's technical data. Typically, the manufacturer's overhaul or maintenance manual would be used.
REFERENCE: AC65-12A, Page 428
AC43.13-1A, Page 267

4232. ANSWER #1

Answer #1 applies to the engine identified because it gives the model number as IVO-355, the serial numbers include the number of our engine, and the hours of time in service apply to our engine. Our engine has 2100 hours of total time in service. The number of hours it has since rebuilding does not matter in relation to this question.
REFERENCE: AC39-7A

4233. ANSWER #1

Information about specific time or cycle limitations for components or parts used on a turbine engine installed in a specific aircraft would be contained in the instructions for continued airworthiness issued by the aircraft manufacturer.
REFERENCE: FAR 91.169 (e)

4234. ANSWER #1

The discharge nozzles in a fuel injected engine have their flow range identified by a letter stamped on the hex of the nozzle body.
REFERENCE: ITP POWERPLANT, Chapter 6, Page 39
AC65-12A, Page 142

4235. ANSWER #1

Electronic fuel flowmeters (vane-type or mass-flow type) are what many modern aircraft are using to measure the amount of fuel consumed by the engine. Fuel flowmeters used with fuel-injected engines are often nothing more than pressure drop indicators. In any case, the measurement shown in the cockpit is in pounds of fuel consumed per hour.
REFERENCE: ITP POWERPLANT, Chapter 5, Page 40
AC65-12A, Page 432

4236. ANSWER #4

Fuel-flow transmitters are usually located in the fuel line between the engine fuel pump and the carburetor.
REFERENCE: ITP POWERPLANT, Chapter 5, Page 40

4237. ANSWER #2
In modern day airplanes, the fuel-flow indicator normally operates on an autosyn system principle. These systems operate on the principle of synchronous movement. When the movement of fuel causes a magnet to rotate in the transmitter, a magnet in the indicator moves the same amount.
REFERENCE: ITP POWERPLANT, Chapter 5, Page 40

4238. ANSWER #3
Autosyn systems operate on alternating current.
REFERENCE: ITP POWERPLANT, Chapter 5, Page 41 (Figure 4B-10)

4239. ANSWER #2
Fuel-flow transmitters are electrical transmitting devices. The signal gets from the transmitter to the indicator electrically.
REFERENCE: ITP POWERPLANT, Chapter 5, Page 41
                        AC65-15A, Page 499

4240. ANSWER #1
Fuel-flow transmitters send an electrical signal to a receiver, which is a fuel-flow gage in the cockpit, to indicate the amount of fuel being consumed in pounds per hour.
REFERENCE: ITP POWERPLANT, Chapter 5, Page 41
                        AC65-15A, Pages 498 and 499

4241. ANSWER #4
Electronic fuel flowmeters (vane-type or mass-flow type) are what many modern aircraft are using to measure the amount of fuel consumed by the engine. Fuel flowmeters used with fuel-injected engines are often nothing more than pressure drop indicators. In any case, the measurement shown in the cockpit is in pounds of fuel consumed per hour.
REFERENCE: ITP POWERPLANT, Chapter 5, Page 40
                        AC65-12A, Page 432
                        AC65-15A, Pages 498 and 499

4242. ANSWER #2
In a ratiometer-type oil temperature indicator system, the gage in the cockpit shows higher temperatures as the resistance in the temperature bulb increases. If the temperature bulb circuit has an open in it, this will produce infinite resistance which will send the temperature gage needle right off the scale.
REFERENCE: AC65-15A, Page 497

4243. ANSWER #4
Fuel-flow indicators operate with a synchronous system. As a magnet moves in the transmitter as a result of fuel flow, a magnet moves in the indicator the same amount. The needle in the indicator is driven magnetically, by what is known as a magnetic linkage.
REFERENCE: ITP POWERPLANT, Chapter 5, Page 41
                        AC65-15A, Page 498

4244. ANSWER #2
The fuel flow indication used for fuel-injected engines is actually a measure of the pressure drop across the fuel-injection nozzles. This is not an especially accurate way of measuring the fuel flow, since a restricted injector nozzle decreases the fuel flow, but it causes an increased pressure drop. This condition will show as an increase in the fuel flow, which is exactly the opposite of what is really happening.
REFERENCE: ITP POWERPLANT, Chapter 5, Page 40

4245. ANSWER #4
The engine analyzer is an adaptation of the laboratory oscilloscope. It is a portable or permanently installed instrument, whose function is to detect, locate, and identify engine operating abnormalities such as those caused by a faulty ignition system, detonation, sticking valves, poor fuel injection, or the like.
REFERENCE: AC65-12A, Page 228

4246. ANSWER #4
Manifold pressure is the pressure in the induction system of a piston engine. It is measured by an absolute pressure gage.
REFERENCE: ITP POWERPLANT, Chapter 5, Page 22
                        AC65-12A, Page 432

4247. ANSWER #3
A manifold pressure gage is, in essence, a barometer. If it becomes disconnected from the engine's induction system outside of the aircraft, the gage will read atmospheric pressure. If it becomes disconnected at the gage (inside the aircraft), the gage will read cabin pressure.
REFERENCE: AIRCRAFT POWERPLANTS, Page 316

4248. ANSWER #2
The purpose of an exhaust gas analyzer is to indicate the fuel/air ratio being burned in the cylinders. It can identify cylinders running too rich or too lean, and can be used to fine tune the fuel metering system.
REFERENCE: THE AVIATION MECHANIC'S MANUAL, Pages 275 and 276

4249. ANSWER #4
The thermocouples used to measure the exhaust gas temperature of turbine engines make use of the dissimilar metals chromel and alumel. When heated, these two metals produce a milliamp current flow which powers a temperature gage.
REFERENCE: ITP POWERPLANT, Chapter 5, Page 30
             AC65-15A, Page 495

4250. ANSWER #2
Tachometer systems, comprised of a tach generator and tach indicator, make use of synchronous motors to send and receive the RPM signal. The indicator's motor turns at the same speed as the tach generator. The shaft of the indicator's motor drives a magnet assembly that causes an aluminum magnetic drag disc to rotate against the restraint of a calibrated hairspring.
REFERENCE: ITP POWERPLANT, Chapter 5, Page 38
             AC65-15A, Page 489

4251. ANSWER #3
When a thermocouple temperature indicating system is used on the cylinder head of a recip engine, the thermocouple's hot junction is placed at the engine. The system's cold junction is at the cylinder head temperature gage.
REFERENCE: ITP POWERPLANT, Chapter 5, Page 30
             AC65-15A, Page 494

4252. ANSWER #3
The typical tachometer system is a three-phase a.c. generator coupled to the aircraft engine, and connected electrically to an indicator mounted on the instrument panel. The generator transmits three-phase power to the synchronous motor in the indicator. The frequency of the transmitted power is proportional to the engine speed.
REFERENCE: ITP POWERPLANT, Chapter 5, Page 37
             AC65-15A, Page 489

4253. ANSWER #2
Because a thermocouple produces its own milliamp current flow, a temperature indicating system using thermocouples does not require any external power. Most systems, however, do use external power and an amplifier to improve the response.
REFERENCE: ITP POWERPLANT, Chapter 5, Page 30
             AC65-15A, Page 494

4254. ANSWER #2
A thermocouple is a circuit or connection of two unlike metals; such a circuit has two junctions. If one of the junctions is heated to a higher temperature than the other, an electromotive force is produced in the circuit. By including a galvanometer in the circuit, this force can be measured. The hotter the high-temperature junction becomes, the greater the electromotive force produced.
REFERENCE: ITP POWERPLANT, Chapter 5, Page 30
             AC65-15A, Page 494

4255. ANSWER #3
A thermocouple is a circuit or connection of two unlike metals; such a circuit has two junctions. If one of the junctions is heated to a higher temperature than the other, an electromotive force is produced in the circuit. By including a galvanometer in the circuit, this force can be measured. The hotter the high-temperature junction becomes, the greater the electromotive force produced.
REFERENCE: ITP POWERPLANT, Chapter 5, Page 30
             AC65-15A, Page 494

4256. ANSWER #1
On recip and turbine powered aircraft, the tachometer is a primary engine instrument. On some turboprop and turboshaft powered aircraft, there is a torque meter gage which is also a primary engine instrument for those aircraft.
REFERENCE: AIRCRAFT ELECTRICITY AND ELECTRONICS, Page 252

4257. ANSWER #1
A manifold pressure gage is, in essence, a barometer. If it becomes disconnected from the engine's induction system outside of the aircraft, the gage will read atmospheric pressure. If it becomes disconnected at the gage (inside the aircraft), the gage will read cabin pressure.
REFERENCE: AIRCRAFT POWERPLANTS, Page 316

4258. ANSWER #2
In dry-sump lubricating systems, the oil temperature bulb may be anywhere in the oil inlet line between the supply tank and the engine. Oil systems for wet-sump engines have the temperature bulb located where it senses oil temperature after the oil passes through the oil cooler. In either case, the oil temperature is being measured before it enters the engine's hot section.
REFERENCE: ITP POWERPLANT, Chapter 8, Page 15
AC65-12A, Page 290

4259. ANSWER #3
Helicopters require a minimum of two tachometers to monitor both the RPM of the engine and of the helicopter rotor system.
REFERENCE: ITP POWERPLANT, Chapter 5, Page 38
AC65-15A, Page 490

4260. ANSWER #3
A thermocouple type of temperature indicating system produces a milliamp current flow when the hot junction is heated to a greater temperature than the cold junction. If the leads to the temperature gage are reversed, the reading will be reversed and the needle will peg out at zero.
REFERENCE: EA-TEP, Page 120
AC65-15A, Page 494

4261. ANSWER #1
Heat sensitive elements and heat sensitive bulb resistance is checked with a wheatstone-bridge meter.
REFERENCE: AC65-15A, Page 493

4262. ANSWER #2
A thermocouple is a circuit or connection of two unlike metals; such a circuit has two junctions. If one of the junctions is heated to a higher temperature than the other, an electromotive force is produced in the circuit. By including a galvanometer in the circuit, this force can be measured. The hotter the high-temperature junction becomes, the greater the electromotive force produced.
REFERENCE: ITP POWERPLANT, Chapter 5, Page 30
AC65-15A, Page 494

4263. ANSWER #3
Engine instruments are usually marked on their face or the face of the glass with colored radials or lines to indicate minimum and maximum ranges. Instruments are not color-coded to pictorially present operating data.
REFERENCE: ITP POWERPLANT, Chapter 5, Page 48
ITP AIRFRAME, Chapter 8, Page 43
AC65-15A, Page 470

4264. ANSWER #3
Thermocouple leads are designed to provide a definite amount of resistance in the thermocouple circuit. Thus, their length or cross-sectional size cannot be altered unless some compensation is made for the change in total resistance. A thermocouple system for a typical turbine engine, for example, has 8 ohms of resistance.
REFERENCE: ITP POWERPLANT, Chapter 5, Page 30
AC65-15A, Page 494

4265. ANSWER #3
Engine pressure ratio in a gas turbine engine is a ratio of the total pressure leaving the turbine (Pt7) to the total pressure entering the engine (Pt2). The ratio of these pressures is an indication of the thrust being produced by the engine. Statement #1 in this question does not give the parts of the ratio in the proper order. The way the statement is worded, it would be a ratio of Pt2 to Pt7. It is actually Pt7 to Pt2. I am still picking the statement as being true because of the way ratios have been stated in other questions. I believe this statement is intended to be true.
REFERENCE: ITP POWERPLANT, Chapter 5, Page 26

4266. ANSWER #1
The typical tachometer system is a three-phase a.c. generator coupled to the aircraft engine, and connected electrically to an indicator mounted on the instrument panel.
REFERENCE: ITP POWERPLANT, Chapter 5, Page 37
                         AC65-15A, Page 489

4267. ANSWER #3
Depending on the installation, there are two instruments used to indicate the thrust being developed by a jet engine. One is the engine tachometer and the other is the engine pressure ratio gage. On a Boeing 727 the EPR gage is the primary indicator of engine thrust and the tachometer is the back-up indicator. On a DC-10, the N1 tach is the primary indicator of thrust and the EPR gage is the back-up indicator.
REFERENCE: ITP POWERPLANT, Chapter 5, Page 26
                         AC65-12A, Page 485

4268. ANSWER #3
Engine pressure ratio is the ratio of the total pressure leaving the turbine to the total pressure entering the engine. It can be expressed as Pt7 ÷ Pt2.
REFERENCE: ITP POWERPLANT, Chapter 5, Page 26

4269. ANSWER #2
The thermocouples used in turbine engines are usually constructed of chromel, a nickel/chromium alloy, and alumel, a nickel/aluminum alloy. These are dissimilar metals which produce a milliamp current flow when heated.
REFERENCE: ITP POWERPLANT, Chapter 5, Page 30
                         AC65-15A, Page 495

4270. ANSWER #1
If a problem with an engine instrument involves the pointer being loose on the shaft, the glass being cracked, the instrument not zeroing out, or the glass being fogged, the instrument must be replaced. A mechanic is not allowed to correct any of these discrepancies.
REFERENCE: AC65-15A, Page 471

4271. ANSWER #1
A Bourdon-tube instrument can be used to measure pressure and temperature. The oil temperature in most small engines is measured by sealing the vapors of methyl chloride in a bourdon tube. As the vapors in the tube are heated and expand, the tube causes a pointer to move which indicates the oil temperature. Oil pressure can also be measured by using a bourdon tube type instrument.
REFERENCE: ITP POWERPLANT, Chapter 5, Pages 24 and 28

4272. ANSWER #1
If an engine instrument has a limit line missing on the glass, a mounting screw which is loose, chipped paint on the case, or a leak at a "B" nut, any or all of these discrepancies can be corrected by a certified mechanic.
REFERENCE: AC65-15A, Page 471

4273. ANSWER #4
If the paint is chipped on an instrument case, no immediate corrective action is required. The only real problem is a cosmetic one.
REFERENCE: AC65-15A, Page 471

4274. ANSWER #2
A manifold pressure gage is, in essence, a barometer. If it becomes disconnected from the engine's induction system outside of the aircraft, the gage will read atmospheric pressure. If it becomes disconnected at the gage (inside the aircraft), the gage will read cabin pressure.
REFERENCE: AIRCRAFT POWERPLANTS, Page 316

4275. ANSWER #4
The mean effective pressure is the average pressure produced in the combustion chamber during the operating cycle. The higher the manifold pressure an engine has, the higher the mean effective pressure will be. In addition, the engine will be more powerful.
REFERENCE: AC65-12A, Page 35

4276. ANSWER #2
The temperature at the turbine should always be monitored when starting a gas turbine engine. The gage in the cockpit might indicate turbine inlet temperature, exhaust gas temperature, or interstage gas temperature. Regardless of the temperature location, the important thing is to monitor the temperature of the turbine.
REFERENCE: ITP POWERPLANT, Chapter 5, Page 33
             AC65-15A, Pages 494 and 495

4277. ANSWER #2
There are a number of places where the oil temperature can be taken on a turbine engine. In a hot-tank system, where the oil cooler is downstream of the pressure pump, the oil temperature is taken as the oil leaves the cooler. At this point, the oil is ready to enter the engine. The AC65-12A states that the oil temperature gage connection is usually located in the pressure inlet to the engine. The pressure inlet to the engine means downstream of the pump.
REFERENCE: ITP POWERPLANT, Chapter 8, Page 25
             AC65-12A, Page 307

4278. ANSWER #1
The anti-icing system on a gas turbine engine typically uses engine bleed air to do the job. This means air from the compressor is being taken away, at an elevated temperature, and routed back to the engine inlet. After the air has done its job of anti-icing, it is dumped back into the inlet. Air being taken away from the compressor, in addition to it being dumped back into the inlet, will reduce the engine pressure ratio.
REFERENCE: ITP POWERPLANT, Chapter 5, Page 26
             ITP POWERPLANT, Chapter 7, Page 24

4279. ANSWER #3
Engine pressure ratio is the ratio of the total pressure leaving the turbine to the total pressure entering the engine. It can be expressed as Pt7 ÷ Pt2.
REFERENCE: ITP POWERPLANT, Chapter 5, Page 26

4280. ANSWER #4
A possible cause of high EGT, high fuel flow, and low RPM in a gas turbine engine is turbine section damage or loss of turbine efficiency. The purpose of the turbine is to extract energy from the gases coming off the combustor and convert that energy into rotary motion to drive the compressor. If the turbine is damaged, it won't pick off as much energy and the RPM will be low. Less energy being picked off by the turbine means more heat energy going out the tailpipe, and therefore higher EGT. The power lever setting will be telling the fuel control to produce a certain engine RPM. Because the turbine is not doing its job, the fuel control will try to compensate by adding more fuel to get the engine RPM up. This will produce higher than normal fuel flow and further add to the high EGT.
REFERENCE: AC65-12A, Page 491

4281. ANSWER #4
Gas turbine engine tachometers are usually calibrated in percent RPM.
REFERENCE: ITP POWERPLANT, Chapter 5, Page 38
             AC65-12A, Page 486
             AC65-15A, Page 488

4282. ANSWER #3
A tachometer on an axial-flow compressor turbine engine is used to monitor the engine during starting and during possible overspeed conditions. On some turbine engines, such as the CF-6 used on the DC-10, the tachometer is used as the primary indicator of thrust.
REFERENCE: AC65-12A, Page 486

4283. ANSWER #1
The EPR indicator is, for the majority of turbine powered airplanes, the primary indicator of engine thrust. Because it is a ratio of the total pressure aft of the turbines to the total pressure at the engine inlet, it is a direct indication of how much energy is left after the turbines have picked off their share to power the engine.
REFERENCE: ITP POWERPLANT, Chapter 5, Page 26
             AC65-12A, Page 485

4284. ANSWER #4
Although the exhaust gas temperature and the turbine inlet temperature are certainly not the same on an operating engine, EGT is a good indicator of what the TIT must be. The engineers who design an engine know how much heat energy the turbine section will absorb from the gases flowing through it. By knowing this, TIT can be calculated as a function of EGT. One monkey wrench that can be thrown into this calculation is damage to the turbine, which will leave it absorbing less heat and therefore produce a false high EGT reading.
REFERENCE: ITP POWERPLANT, Chapter 5, Page 33
            AC65-12A, Page 486

4285. ANSWER #4
The EPR indicator is, for the majority of turbine powered airplanes, the primary indicator of engine thrust. Because it is a ratio of the total pressure aft of the turbines to the total pressure at the engine inlet, it is a direct indication of how much energy is left after the turbines have picked off their share to power the engine.
REFERENCE: ITP POWERPLANT, Chapter 5, Page 26
            AC65-12A, Page 485

4286. ANSWER #4
In a gas turbine engine, the turbine discharge pressure sensor is located immediately aft of the last turbine stage.
REFERENCE: AC65-12A, Page 485

4287. ANSWER #1
Gas turbine engine tachometers are usually calibrated in percent RPM.
REFERENCE: ITP POWERPLANT, Chapter 5, Page 38
            AC65-12A, Page 486
            AC65-15A, Page 488

4288. ANSWER #4
To detect fires or overheat conditions, detectors are placed in the various zones to be monitored. Fires are detected in reciprocating engine aircraft by using one or more of the following:
1. Overheat detectors.
2. Rate-of-temperature-rise detectors.
3. Flame detectors.
4. Observation by crewmembers.
REFERENCE: ITP POWERPLANT, Chapter 5, Page 53
            AC65-12A, Page 391

4289. ANSWER #3
The function of a fire detection system is to activate a warning device in the event of a powerplant fire.
REFERENCE: ITP POWERPLANT, Chapter 5, Page 51
            AC65-12A, Page 391

4290. ANSWER #3
It is necessary to interpret the chart to answer this question. A minus 15 degrees Fahrenheit is half-way between the 10 and 20. Halfway between their minimum pressures is 115 psig and halfway between their maximum pressures is 199 psig. So the pressure limits would be 115 minimum to 199 maximum.
REFERENCE: AC65-12A, Page 406

4291. ANSWER #1
In a turbine engine powered aircraft, the fire extinguishing portion of a complete fire protection system typically includes a cylinder or container of an extinguishing agent for each engine and nacelle area. The container of agent is normally equipped with two discharge valves that are operated by electrically discharged cartridges. The two valves are the main and the reserve controls that release and route the agent to the pod and pylon in which the container is located or to the other engine on the same wing.
REFERENCE: ITP POWERPLANT, Chapter 5, Page 59
            AC65-12A, Pages 400 and 401

4292. ANSWER #4
When a CO_2 fire extinguisher container is discharged, the compressed liquid flows in one rapid burst to the outlets in the distribution line of the affected engine. Contact with the air converts the liquid into gas and "snow" which smothers the flame.
REFERENCE: AC65-12A, Page 396

4293. ANSWER #1

A typical high rate of discharge container is equipped with a pressure gage, a discharge plug, and a safety discharge connection. The discharge plug is sealed with a breakable disk combined with an explosive charge that is electrically detonated to discharge the contents of the bottle. The safety discharge connection, or fusible disk, is capped at the inboard side of the engine strut with a red indication disk. If the temperature rises beyond a predetermined safe value, the disk will rupture, dumping the agent overboard.
REFERENCE: AC65-12A, Page 401

4294. ANSWER #2

A continuous-loop detector or sensing system permits more complete coverage of a fire hazard area than any of the spot-type temperature detectors. Continuous-loop systems are versions of the thermal switch system. They are overheat systems, heat-sensitive units that complete electrical circuits at a certain temperature.
REFERENCE: ITP POWERPLANT, Chapter 5, Pages 53 and 54
             AC65-12A, Page 393

4295. ANSWER #4

Spot detector systems operate on a different principle from the continuous loop. Each detector unit consists of a bimetallic thermoswitch, electrically above ground potential, which closes when heated to a high temperature. The closing of the switch completes a circuit which sets off the fire alarm.
REFERENCE: ITP POWERPLANT, Chapter 5, Page 51
             AC65-12A, Page 393

4296. ANSWER #4

In a typical engine fire extinguishing system, the extinguishing agent is distributed through spray nozzles and perforated tubing. The perforated tubing distribution system is more common with recip engines and the spray nozzles are most often used with turbine engines.
REFERENCE: AC65-12A, Pages 396 and 402
             AC65-15A, Page 417

4297. ANSWER #1

From a standpoint of toxicity and corrosion hazards, carbon dioxide is the safest agent to use. It was for many years the most widely used agent. Now halogenated hydrocarbons are the most widely used agents.
REFERENCE: AC65-12A, Page 395

4298. ANSWER #1

To detect fires or overheat conditions, detectors are placed in the various zones to be monitored. Fires are detected in reciprocating engine aircraft by using one or more of the following:
1. Overheat detectors.
2. Rate-of-temperature-rise detectors.
3. Flame detectors.
4. Observation by crewmembers.
REFERENCE: ITP POWERPLANT, Chapter 5, Page 53
             AC65-12A, Page 391

4299. ANSWER #2

In the Kidde continuous loop detector systems, there is an inconel tube with two wires running through it. One of the wires has potential to it and the other is a source to ground. In the Fenwal system, there is an inconel tube with one wire running through it. The wire has potential to it and the tube is the source to ground. In both systems, the potential and ground are separated by a core material which, when cold, acts as a resistor. When the core material is heated to a specified temperature, however, it acts as a conductor and allows the potential to find a path to ground. When this circuit is completed, it causes a fire alarm to sound.
REFERENCE: ITP POWERPLANT, Chapter 5, Page 54
             AC65-12A, Page 393

4300. ANSWER #1

Carbon dioxide is the most satisfactory agent to use for a carburetor or intake fire. Whatever agent is used will be drawn through the carburetor and into the engine. CO_2, unlike many of the other agents, will not damage the carburetor or the engine.
REFERENCE: AC65-9A, Page 491

4301. ANSWER #1

The service life of fire extinguisher discharge cartridges is calculated from the manufacturer's date stamp, which is usually placed on the face of the cartridge. The cartridge service life recommended by the manufacturer is usually in terms of hours below a predetermined temperature limit.
REFERENCE: AC65-12A, Page 406
             AC65-15A, Page 426

4302. ANSWER #4
The Fenwal fire-detection system is wired in parallel between two separate circuits so that a short in either leg of the system will not cause a false fire warning. The system is wired so that one leg of the circuit supplies potential to the detectors and the other leg serves as a path to ground. If the ground leg of the circuit has a short in it, a false fire warning will not occur because this portion of the circuit is already grounded. If the powered leg of the circuit has a short in it, the current flow will be great enough to trip the relay in the circuit. The tripped relay causes the powered leg to become the grounded part of the system and the grounded leg to become powered. In either case, the short has not caused a false fire warning.
REFERENCE: EA-TEP, Pages 134 and 135
　　　　　　　　 AC65-12A, Pages 393 and 394

4303. ANSWER #2
The thermocouple fire warning system operates on an entirely different principle than the thermal switch system. A thermocouple depends upon the rate of temperature rise and will not give a warning when an engine slowly overheats or a short circuit develops. In each thermocouple, there is a hot junction and a cold or reference junciton. If both of these junctions heat up at the same rate, no fire warning will be given regardless of the temperature.
REFERENCE: ITP POWERPLANT, Chapter 5, Page 53
　　　　　　　　 AC65-12A, Page 392

4304. ANSWER #2
When the warning light comes on in the cockpit indicating fire, in a Boeing 727 for example, the pilot pulls the fire handle. This arms the fire bottle discharge switch (which is exposed now that the fire handle has been pulled), trips the generator field relay, shuts off the fuel and hydraulic fluid to the engine, shuts off the engine bleed air, and deactivates the engine-driven hydraulic pump low-pressure lights.
REFERENCE: ITP POWERPLANT, Chapter 5, Page 57

4305. ANSWER #3
The Lindberg continuous-element fire detection system operates on a pneumatic detection principle. A stainless steel tube is filled with an inert gas, typically helium. The principle of operation is based on the laws of gases. If the volume of the gas is held constant, its pressure will increase as temperature increases. Thus the helium gas between the two tubing walls will exert a pressure proportional to the average absolute temperature along the entire length of the tube. One end of the tube is connected to a small chamber containing a metal diaphragm switch. At a specific temperature, the pressure of the gas is great enough to activate the switch and set off a fire alarm.
REFERENCE: AIRCRAFT POWERPLANTS, Pages 324 and 325

4306. ANSWER #2
The Lindberg continuous-element fire detection system operates on a pneumatic detection principle. A stainless steel tube is filled with an inert gas, typically helium. The principle of operation is based on the laws of gases. If the volume of the gas is held constant, its pressure will increase as temperature increases. Thus the helium gas between the two tubing walls will exert a pressure proportional to the average absolute temperature along the entire length of the tube. One end of the tube is connected to a small chamber containing a metal diaphragm switch. At a specific temperature, the pressure of the gas is great enough to activate the switch and set off a fire alarm.
REFERENCE: AC65-15A, Page 411

4307. ANSWER #1
In the Kidde continuous loop detector systems, there is an inconel tube with two wires running through it. One of the wires has potential to it and the other is a source to ground. In the Fenwal system, there is an inconel tube with one wire running through it. The wire has potential to it and the tube is the source to ground. In both systems, the potential and ground are separated by a core material which, when cold, acts as a resistor. When the core material is heated to a specified temperature, however, it acts as a conductor and allows the potential to find a path to ground. When this circuit is completed, it causes a fire alarm to sound.
REFERENCE: ITP POWERPLANT, Chapter 5, Page 54
　　　　　　　　 AC65-12A, Page 393

4308. ANSWER #4
In the Kidde continuous loop detector systems, there is an inconel tube with two wires running through it. One of the wires has potential to it and the other is a source to ground. In the Fenwal system, there is an inconel tube with one wire running through it. The wire has potential to it and the tube is the source to ground. In both systems, the potential and ground are separated by a core material which, when cold, acts as a resistor. When the core material is heated to a specified temperature, however, it acts as a conductor and allows the potential to find a path to ground. When this circuit is completed, it causes a fire alarm to sound.
REFERENCE: ITP POWERPLANT, Chapter 5, Page 54
　　　　　　　　 AC65-12A, Page 393

4309. ANSWER #2

Class "B" fires are defined as fires in flammable petroleum products or other flammable or combustible liquids, greases, solvents, paints, etc.
REFERENCE: ITP POWERPLANT, Chapter 5, Page 56
                        AC65-15A, Page 411

4310. ANSWER #2

The thermocouple fire warning system operates on an entirely different principle than the thermal switch system. A thermocouple depends upon the rate of temperature rise and will not give a warning when an engine slowly overheats or a short circuit develops. In each thermocouple, there is a hot junction and a cold or reference junction. If both of these junctions heat up at the same rate, no fire warning will be given regardless of the temperature.
REFERENCE: ITP POWERPLANT, Chapter 5, Page 53
                        AC65-12A, Page 392

4311. ANSWER #4

Class "C" fires are defined as fires involving energized electrical equipment where the electrical non-conductivity of the extinguishing media is of importance.
REFERENCE: ITP POWERPLANT, Chapter 5, Page 56
                        AC65-15A, Page 411

4312. ANSWER #2

Kidde and Fenwal both make what are called continuous-loop fire detection systems. Although they can experience a break and still give a fire warning, their circuits must be complete for the press-to-test operation to work.
REFERENCE: ITP POWERPLANT, Chapter 5, Pages 53 and 54
                        AC65-12A, Page 393
                        AC65-15A, Page 410

4313. ANSWER #3

A high pressure bottle stamped "ICC 1-70" or "DOT 1-70" indicates that the bottle is approved by the Interstate Commerce Commission or the Department of Transportation, and that it was hydrostatically checked in January of 1970. In terms of approval and testing, the same guidelines apply to fire extinguisher bottles that apply to oxygen system bottles.
REFERENCE: EA-AOS-1, Page 19

4314. ANSWER #3

Fire extinguishing systems typically have two disks which are used to indicate that the agent has been emptied. The yellow disk blows when the agent has been emptied by a normal discharge and the red disk blows when the agent has been emptied because of an over-pressure.
REFERENCE: ITP POWERPLANT, Chapter 5, Page 59
                        AC65-15A, Page 418

4315. ANSWER #3

Carbon dioxide has been used for many years to extinguish flammable fluid fires and fires involving electrical equipment.
REFERENCE: AC65-15A, Page 415

4316. ANSWER #3

Kidde and Fenwal both make what are called continuous-loop fire detection systems. Although they can experience a break and still give a fire warning, their circuits must be complete for the press-to-test operation to work.
REFERENCE: ITP POWERPLANT, Chapter 5, Pages 53 and 54
                        AC65-12A, Page 393
                        AC65-15A, Page 410

4317. ANSWER #1

When the systems are being tested, the thermocouple and the Lindberg fire detection systems heat up their detectors to simulate a fire condition.
REFERENCE: AC65-15A, Pages 409 to 411

4318. ANSWER #3

By replacing the slip rings of the basic a.c. generator with two half-cylinders, called a commutator, a basic d.c. generator is obtained. As the generator's armature rotates, the commutator elements are actually acting as a switch. Though the current actually reverses its direction in the loop in exactly the same way as in the a.c. generator, the commutator action causes the current to flow always in the same direction through the external circuit.
REFERENCE: AC65-9A, Pages 388 and 399

4319. ANSWER #3
Because of the low resistance in the windings, the series motor is able to draw a large current in starting. This starting current, in passing through both the field and armature windings, produces a high starting torque, which is the series motor's principal advantage.
REFERENCE: AC65-9A, Page 446

4320. ANSWER #4
The stationary field strength in a direct-current generator is varied according to the load requirements by short circuiting the field rheostat for varying periods of time. As the load on the generator increases, the field rheostat is shorted for a longer period of time, increasing the current flowing through the stationary field and increasing the output of the generator.
REFERENCE: ITP AIRFRAME, Chapter 5, Page 10
             AC65-9A, Pages 398 and 399

4321. ANSWER #2
The typical starter motor is a 12 or 24 volt direct current series-wound motor, which develops high starting torque.
REFERENCE: ITP POWERPLANT, Chapter 7, Page 44
             AC65-12A, Page 265

4322. ANSWER #4
The frequency of the alternator voltage depends upon the speed of rotation of the rotor and the number of poles. The faster the speed, the higher the frequency will be. The more poles on the rotor, the higher the frequency will be for a given speed.
REFERENCE: AC65-9A, Page 417

4323. ANSWER #4
A motor connected to a battery may draw a fairly high current on starting, but as the armature speed increases, the current flowing through the armature decreases. When the armature in a motor rotates in a magnetic field, a voltage is induced in its windings. This voltage is called the back or counter electromotive force (e.m.f.) and is opposite in direction to the voltage applied to the motor from the external source. Counter e.m.f. opposes the current which causes the armature to rotate. The current flowing through the armature, therefore, decreases as the counter e.m.f. increases.
REFERENCE: ITP AIRFRAME, Chapter 5, Page 53
             AC65-9A, Pages 447 and 448

4324. ANSWER #3
Alternators are not always connected directly to the airplane engine like d.c. generators. Since the various electrical devices operating on a.c. supplied by alternators are designed to operate at a certain voltage and at a specified frequency, the speed of the alternators must be constant. To accomodate this need, some alternators are driven by the engine through a constant-speed drive.
REFERENCE: ITP AIRFRAME, Chapter 5, Page 15
             AC65-9A, Page 423

4325. ANSWER #2
A rough or pitted commutator should be smoothed out using a very fine sandpaper, such as 000, and then cleaned and polished with a clean, dry cloth.
REFERENCE: AC65-9A, Page 434

4326. ANSWER #4
Generator circuits in aircraft are supplied with switches which can be used to energize or de-energize the generator. In the case of a malfunction, the generator circuit should be de-energized.
REFERENCE: AC65-9A, Page 407

4327. ANSWER #3
In a generator circuit using a vibrating-type voltage regulator, it is the opening and closing of a set of points which control the output of the generator. The longer the points stay closed, the greater the output of the generator. If the points stick closed, the generator's output will increase to its maximum.
REFERENCE: ITP AIRFRAME, Chapter 5, Page 12
             AC65-9A, Page 399

4328. ANSWER #4
Alternators are not always connected directly to the airplane engine like d.c. generators. Since the various electrical devices operating on a.c. supplied by alternators are designed to operate at a certain voltage and at a specified frequency, the speed of the alternators must be constant. To accomodate this need, some alternators are driven by the engine through a constant-speed drive.
REFERENCE: ITP AIRFRAME, Chapter 5, Page 15
             AC65-9A, Page 423

4329. ANSWER #2
In a generator circuit, the voltage regulator controls the output of the generator by varying the resistance, and therefore the current flow, in the field circuit. A short circuit between the positive armature lead and the field lead will effectively take the resistance of the voltage regulator out of the circuit. This will increase the current flow through the field circuit and increase the voltage output of the generator.
REFERENCE: AC65-9A, Pages 398 and 399 (Figure 9-22)

4330. ANSWER #4
According to FAR 25, aircraft that operate more than one generator connected to a common electrical system must be provided with individual generator switches that can be operated from the cockpit during flight.
REFERENCE: FAR 25.1351 (B) (5)

4331. ANSWER #1
Efficient operation of electrical equipment in an aircraft depends on a constant voltage supply from the generator. Among the factors which determine the voltage output of a generator, only one, the strength of the field current, can be conveniently controlled.
REFERENCE: ITP AIRFRAME, Chapter 5, Page 10
                       AC65-9A, Page 398

4332. ANSWER #2
The typical starter motor is a 12 or 24 volt direct current series-wound motor, which develops high starting torque.
REFERENCE: ITP POWERPLANT, Chapter 7, Page 44
                       AC65-12A, Page 265

4333. ANSWER #3
Generators are rated for a specified number of amps output at a set voltage. As the load on a generator increases, the amperage output will increase up to the generator's limit, while the voltage remains constant.
REFERENCE: AC65-9A, Page 398

4334. ANSWER #4
It is learned in the study of physics that work and power cannot be gained, meaning you cannot get more out than you put in. The same is true of the generator in the aircraft. As the output of the generator increases, the force supplied by the engine to turn it must increase.
REFERENCE: AC65-15A, Page 458

4335. ANSWER #3
The problem of voltage regulation in an a.c. system does not differ basically from that in a d.c. system. In each case the function of the regulator system is to control voltage, maintain a balance of circulating current throughout the system, and eliminate sudden changes in voltage when a load is applied to the system.
REFERENCE: ITP AIRFRAME, Chapter 5, Page 13
                       AC65-9A, Page 417

4336. ANSWER #4
The purpose of the reverse-current cutout relay is to automatically disconnect the battery from the generator when the generator voltage is less than the battery voltage. If this device were not used in the generator circuit, the battery would discharge through the generator and try to drive it as a motor.
REFERENCE: ITP AIRFRAME, Chapter 5, Pages 12 and 13
                       AC65-9A, Page 401

4337. ANSWER #3
When the coils of an armature become short circuited, the operating temperature increases substantially. The increased temperature causes the solder which separates the segments on the commutator to melt. An open armature may be the result, but the problem is caused by a shorted armature.
REFERENCE: AIRCRAFT ELECTRICITY AND ELECTRONICS, Page 121
                       AC65-9A, Page 411

4338. ANSWER #3
In a brushless alternator, voltage is built up by using permanent magnet interpoles in the exciter stator. The permanent magnets assure a voltage buildup, precluding the necessity of field flashing.
REFERENCE: AC65-9A, Page 415

4339. ANSWER #3
The automatic ignition relight system is activated differently on different aircraft. Some turboprop aircraft have a system which is activated by a decrease in torque oil pressure. A popular method of activating the system, as identified in this question, is by a decrease in compressor discharge pressure.
REFERENCE: ITP POWERPLANT, Chapter 3, Page 45

4340. ANSWER #4
In the brush-type alternator, the rotor has an exciter winding as well as the field winding. The exciter field is controlled by the voltage regulator to vary the current produced by the exciter armature. The a.c. produced in this armature is rectified by a brush-and-commutator arrangement, and the resulting d.c. output is fed through slip rings into the field winding of the rotor. By using this arrangement, the needed amount of alternator field current can be controlled by a much smaller current through the voltage regulator.
REFERENCE: ITP AIRFRAME, Chapter 5, Page 15

4341. ANSWER #3
In cold climates, the state of charge in a storage battery should be kept at a maximum. A fully charged battery will not freeze even under the most severe weather conditions, but a discharged battery will freeze very easily.
REFERENCE: ITP GENERAL, Chapter 2, Page 43
                       AIRCRAFT ELECTRICITY AND ELECTRONICS, Page 91

4342. ANSWER #1
The capacity of a storage battery is rated in ampere-hours (amperes furnished by the battery times the amount of time current can be drawn). This rating indicates how long the battery may be used at a given rate before it becomes completely discharged. A battery which can deliver 45 amps for 2.5 hours has a 112.5 ampere-hour capacity ($45 \times 2.5 = 112.5$).
REFERENCE: ITP GENERAL, Chapter 2, Page 43
                       AC65-9A, Page 310

4343. ANSWER #3
The capacity of a storage battery is rated in ampere-hours (amperes furnished by the battery times the amount of time current can be drawn). This rating indicates how long the battery may be used at a given rate before it becomes completely discharged. A battery with a 140 ampere-hour capacity can deliver 15 amps for 9.33 hours ($140 \div 15 = 9.33$).
REFERENCE: ITP GENERAL, Chapter 2, Page 43
                       AC65-9A, Page 310

4344. ANSWER #4
The aircraft starter circuit is not fused because the current draw is so great that the fuse would continually be blowing.
REFERENCE: AC65-12A, Pages 264 and 265

4345. ANSWER #1
Ideally there should be only two terminals attached to a stud, but in no case should more than four terminals be attached.
REFERENCE: ITP AIRFRAME, Chapter 5, Page 38
                       AC43.13-1A, Page 189

4346. ANSWER #1
No more than four bonding jumper wires should be attached to one terminal grounded to a flat surface.
REFERENCE: AC43.13-1A, Page 189

4347. ANSWER #1
As a general rule, starter brushes should be replaced when they are worn down to approximately one-half of their original length. Allowing the wear to progress much farther will affect the spring tension and the ability of the brush to stay in contact with the commutator.
REFERENCE: AC65-12A, Page 270

4348. ANSWER #4
Switches should be derated from their nominal current rating when they are used to control direct current motors. These motors will draw several times their rated current during starting, and magnetic energy stored in their armature and field coils is released when the control switch is opened.
REFERENCE: AC65-12A, Page 262
                       AC43.13-1A, Page 177

4349. ANSWER #4
One of the purposes of bonding and grounding the structure of an aircraft is to provide current return paths. If the bonding and grounding is properly done, there should be virtually no resistance in the return path.
REFERENCE: ITP AIRFRAME, Chapter 5, Page 45
                       AC65-12A, Page 255

4350. ANSWER #4
When specified on the engineering drawing, or when accomplished as a local practice, parallel wires must sometimes be twisted. The following are the most common examples:
1. Wiring in the vicinity of magnetic compass or flux valve.
2. Three-phase distribution wiring.
3. Radio wiring.
REFERENCE: AC65-12A, Page 243

4351. ANSWER #2
Solenoid-operated electrical switches are built so that they are spring loaded open. When the circuit to the solenoid is completed, the solenoid closes a set of contacts and current continues to flow so long as the circuit to the solenoid is complete.
REFERENCE: ITP GENERAL, Chapter 2, Page 24
             ITP AIRFRAME, Chapter 5, Page 47
             AC65-9A, Pages 319 and 320

4352. ANSWER #3
The electrical load check described in this question is an acceptable method, and the total load on the system is well within the generator's capabilities. The total load on the system is only 76% of the generator's capabilities. According to AC43.13-1A, where the use of placards or monitoring devices is not practicable or desired, and where assurance is needed that the battery in a typical small aircraft generator/battery power source will be charged in flight, the total continuous connected electrical load may be held to approximately 80% of the total rated generator output capacity.
REFERENCE: AC65-12A, Page 261
             AC43.13-1A, Page 175

4353. ANSWER #3
When pulling electrical wires or cables through conduit, soapstone talc can be used to dust the cables before they are pulled through. This will keep the wire from binding and chafing against the walls of the conduit.
REFERENCE: AC65-15A, Pages 442 and 443

4354. ANSWER #3
One of the purposes of bonding and grounding the structure of an aircraft is to provide current return paths. If the bonding and grounding is properly done, there should be virtually no resistance in the return path.
REFERENCE: ITP AIRFRAME, Chapter 5, Page 45
             AC65-12A, Page 255

4355. ANSWER #4
A likely cause of a turbine engine not rotating when the starter-generator circuit is energized is a defective starter relay. A defective throttle ignition switch could also keep an engine from rotating, but it would also keep the starter-generator circuit from being energized.
REFERENCE: ITP POWERPLANT, Chapter 7, Page 53
             AC65-12A, Pages 275 and 276

4356. ANSWER #1
Excessive spring pressure will cause rapid wear of the brushes. Too little pressure, however, will allow "bouncing" of the brushes, resulting in burned and pitted surfaces.
REFERENCE: AC65-9A, Pages 406 and 460

4357. ANSWER #3
The voltage drop in the main power wires from the generation source or the battery to the bus should not exceed 2% of the regulated voltage. This applies when the generator is carrying rated current or the battery is being discharged at the 5-minute rate.
REFERENCE: AC43.13-1A, Page 179

4358. ANSWER #4
Hazardous errors in switch operation may be avoided by logical and consistent installation. "On-off" two position switches should be mounted so that the "on" position is reached by an upward or forward movement of the toggle.
REFERENCE: AC43.13-1A, Page 177

4359. ANSWER #3
Switches should be derated from their nominal current rating when they are used to control a direct current motor. This type of motor will draw several times its rated current during starting, so the switch must be capable of handling this without the contacts burning or welding together.
REFERENCE: AC43.13-1A, Page 177

4360. ANSWER #3
An arcing fault between an electric wire and a metallic flammable fluid line may puncture the line and result in a serious fire. Consequently, every effort should be made to avoid this hazard by physical separation of the wire from lines or equipment containing oil, fuel, hydraulic fluid, or alcohol. When separation is impractical, locate the electric wire above the flammable fluid line and securely clamp it to the structure.
REFERENCE: ITP AIRFRAME, Chapter 5, Page 43
                         AC43.13-1A, Page 187

4361. ANSWER #2
The use of this electric wire chart is fully explained in the AC65-15A.
REFERENCE: ITP AIRFRAME, Chapter 5, Pages 34 to 36
                         AC65-15A, Pages 436 to 439

4362. ANSWER #3
The use of this electric wire chart is fully explained in the AC65-15A.
REFERENCE: ITP AIRFRAME, Chapter 5, Pages 34 to 36
                         AC65-15A, Pages 436 to 439

4363. ANSWER #4
The Farad is the basic unit of measure for capacitance.
REFERENCE: ITP GENERAL, Chapter 3, Page 22
                         AC65-9A, Page 348

4364. ANSWER #2
In selecting or substituting a capacitor for use in a particular circuit, the following must be considered; (1) The value of capacitance desired and (2) the amount of voltage to which the capacitor is to be subjected. If the voltage applied across the plates is too great, the dielectric will break down and arcing will occur between the plates.
REFERENCE: ITP GENERAL, Chapter 3, Page 23
                         AC65-9A, Page 351

4365. ANSWER #2
The effective value of alternating current is the same as the value of a direct current which can produce an equal heating effect. The effective value is less than the maximum value, being equal to .707 times the maximum value.
REFERENCE: ITP GENERAL, Chapter 3, Page 10
                         AC65-9A, Page 343

4366. ANSWER #1
The greater the area of the plates in a capacitor, the greater the storage capability. The greater the distance between the plates in a capacitor, the lesser the storage capability.
REFERENCE: ITP GENERAL, Chapter 3, Page 22

4367. ANSWER #4
Capacitors may be combined in parallel or series to give equivalent values, which may be either the sum of the individual values (in parallel) or a value less than that of the smallest capacitance (in series).
REFERENCE: ITP GENERAL, Chapter 3, Page 24
                         AC65-9A, Page 350

4368. ANSWER #4
The resistance of an oil flow is known as its viscosity. An oil which flows slowly is viscous or has a high viscosity. If it flows freely, it has a low viscosity. Unfortunately, the viscosity of oil is affected by temperature. It is not uncommon for some grades of oil to become practically solid in cold weather. This increases drag and makes circulation almost impossible. Other oils may become so thin at high temperature that the oil film is broken, resulting in rapid wear of the moving parts. As the oil thins out, its resistance to flow decreases and so does the oil pressure in the engine. Low oil pressure takes away the ability of the oil to support loads.
REFERENCE: AC65-12A, Page 285

4369. ANSWER #1
Gas-turbine and recip engine oils should never be mixed. Gas-turbine oils are synthetic base and recip oils are mineral base. Synthetic oil in a recip engine would most likely attack the rubber seals and could result in engine failure, unless it was designed for use in that engine.
REFERENCE: ITP POWERPLANT, Chapter 8, Pages 16 and 17

4370. ANSWER #2
Engine-driven vacuum pumps use an air-oil separator. Oil and air in the vacuum pump are exhausted through the separator, which separates the oil from the air; the air is vented overboard and the oil is returned to the engine sump.
REFERENCE: ITP AIRFRAME, Chapter 8, Page 23
AC65-15A, Page 503

4371. ANSWER #3
Generally, commercial aviation oils are classified numerically, such as 80, 100, 140, etc., which are an approximation of their viscosity as measured by a testing instrument called the Saybolt Universal Viscosimeter. In this instrument a tube holds a specific quantity of the oil to be tested. The oil is brought to an exact temperature by a liquid bath surrounding the tube. The time in seconds required for exactly 60 cubic centimeters of oil to flow through an accurately calibrated orifice is recorded as a measure of the oil's viscosity.
REFERENCE: ITP POWERPLANT, Chapter 8, Page 4
AC65-12A, Page 286

4372. ANSWER #3
Viscosity index is a measure of the change in viscosity of an oil with a given change in temperature. The smaller the change in the viscosity for a given temperature change, the higher the viscosity index.
REFERENCE: ITP POWERPLANT, Chapter 8, Page 5

4373. ANSWER #1
Viscosity index is a measure of the change in viscosity of an oil with a given change in temperature. The smaller the change in the viscosity for a given temperature change, the higher the viscosity index.
REFERENCE: ITP POWERPLANT, Chapter 8, Page 5

4374. ANSWER #4
The many requirements for lubricating oils are met in the synthetic oils developed specifically for turbine engines. Synthetic oil has two principal advantages over petroleum oil. It has less tendency to deposit lacquer and coke and less tendency to evaporate at high temperature.
REFERENCE: ITP POWERPLANT, Chapter 8, Page 16
AC65-12A, Page 302

4375. ANSWER #3
The oil used in reciprocating engines has a relatively high viscosity because of:
(1) Large engine operating clearances due to the relatively large size of the moving parts, the different materials used, and the different rates of expansion of the various materials.
(2) The high operating temperatures.
(3) The high bearing pressures.
REFERENCE: AC65-12A, Page 285

4376. ANSWER #1
The oil selected for aircraft engine lubrication must be light enough to circulate freely, yet heavy enough to provide the proper oil film at engine operating temperatures.
REFERENCE: AC65-12A, Page 285

4377. ANSWER #2
In addition to reducing friction, the oil film acts as a cushion between metal parts. This cushioning effect is particularly important for such parts as reciprocating engine crankshafts and connecting rods, which are subject to shock-loading. As oil circulates through the engine, it absorbs heat from the parts. The oil also aids in forming a seal between the piston and the cylinder wall to prevent leakage of the gases from the combustion chamber. Oils also reduce the abrasive wear by picking up foreign particles and carrying them away to a filter. An equally important function of oil is the prevention of corrosion of the metal parts inside an engine.
REFERENCE: ITP POWERPLANT, Chapter 8, Pages 1 and 2
AC65-12A, Page 285

4378. ANSWER #1
The many requirements for lubricating oils are met in the synthetic oils developed specifically for turbine engines. Synthetic oil has two principal advantages over petroleum oil. It has less tendency to deposit lacquer and coke and less tendency to evaporate at high temperature.
REFERENCE: ITP POWERPLANT, Chapter 8, Page 16
AC65-12A, Page 302

4379. ANSWER #3
The flash point of oil is an important characteristic when selecting a lubricant for use in an aircraft engine. Flash point and fire point are determined by laboratory tests that show the temperature at which a liquid will begin to give off ignitable vapors (flash) and the temperature at which there are sufficient vapors to support a flame (fire). An aircraft engine oil should have a high flash point.
REFERENCE: ITP POWERPLANT, Chapter 8, Page 16
AC65-12A, Page 285

4380. ANSWER #1
The resistance of an oil to flow is known as its viscosity.
REFERENCE: ITP POWERPLANT, Chapter 8, Page 16
            AC65-12A, Page 285

4381. ANSWER #1
Both wet and dry sump lubrication systems are used in gas turbine engines. Most turbojet engines are of the axial-flow configuration, and use a dry sump lubrication system. The engine's bearings are pressure lubricated and the gearboxes are pressure and spray lubricated.
REFERENCE: ITP POWERPLANT, Chapter 8, Page 19
            AC65-12A, Page 302

4382. ANSWER #1
In addition to reducing friction, the oil film acts as a cushion between metal parts. This cushioning effect is particularly important for such parts as reciprocating engine crankshafts and connecting rods, which are subject to shock-loading. As oil circulates through the engine, it absorbs heat from the parts. The oil also aids in forming a seal between the piston and the cylinder wall to prevent leakage of the gases from the combustion chamber. Oils also reduce the abrasive wear by picking up foreign particles and carrying them away to a filter. An equally important function of oil is the prevention of corrosion of the metal parts inside an engine.
REFERENCE: ITP POWERPLANT, Chapter 8, Pages 1 and 2
            AC65-12A, Page 285

4383. ANSWER #4
Several factors must be considered in determining the proper grade of oil to use in a particular engine. The operating load, rotational speeds, and operating temperatures are the most important.
REFERENCE: AC65-12A, Page 285

4384. ANSWER #2
One characteristic of oil is its specific gravity. Specific gravity is a comparison of the weight of the substance to the weight of an equal volume of distilled water at a specified temperature.
REFERENCE: AC65-12A, Page 285

4385. ANSWER #1
The viscosity of oil is a measure of its resistance to flow. The factor which affects an oil's viscosity the most is temperature. The higher the temperature, the more an oil will thin out and the less viscous it will become.
REFERENCE: AC65-12A, Page 285

4386. ANSWER #3
In general, lubricants of animal and vegetable origin are chemically unstable at high temperatures, often perform poorly at low temperatures, and are unsuited for aircraft-engine lubrication.
REFERENCE: AIRCRAFT POWERPLANTS, Page 165

4387. ANSWER #1
In general, lubricants of animal and vegetable origin are chemically unstable at high temperatures, often perform poorly at low temperatures, and are unsuited for aircraft-engine lubrication. The lubricants which are best suited for aircraft use, are the mineral base and the synthetic base.
REFERENCE: ITP POWERPLANT, Chapter 8, Pages 2 and 3

4388. ANSWER #3
In addition to reducing friction, the oil film acts as a cushion between metal parts. This cushioning effect is particularly important for such parts as reciprocating engine crankshafts and connecting rods, which are subject to shock-loading. As oil circulates through the engine, it absorbs heat from the parts. The oil also aids in forming a seal between the piston and the cylinder wall to prevent leakage of the gases from the combustion chamber. Oils also reduce the abrasive wear by picking up foreign particles and carrying them away to a filter. An equally important function of oil is the prevention of corrosion of the metal parts inside an engine.
REFERENCE: ITP POWERPLANT, Chapter 8, Pages 1 and 2
            AC65-12A, Page 285

4389. ANSWER #1
The type of oil pump used most often in gas turbine engines is the gear type. The vane type, which is among the possible answers, is also very popular.
REFERENCE: ITP POWERPLANT, Chapter 8, Page 20
            AC65-12A, Page 304

4390. ANSWER #1
Pulling back the throttle too rapidly on an engine which is operating at high temperatures could cause the oil trapped in the piston ring grooves to carbonize. A sudden reduction in engine RPM means a reduction in cooling airflow, which causes the cylinders to momentarily be at a higher than normal temperature. This high temperature can cause the oil in the ring grooves to carbonize.
REFERENCE: AIRCRAFT PROPULSION POWERPLANTS, Page 483

4391. ANSWER #2

Generally, the oil leaving an engine must be cooled before it is re-circulated. Obviously, the amount of cooling must be controlled if the oil is to return to the engine at the correct temperature. The oil temperature regulator, located in the return line to the tank, provides for this controlled cooling.
REFERENCE: AC65-12A, Page 290

4392. ANSWER #2

If the return line between the scavenge pump and the oil cooler separates on an engine, the return oil will be pumped overboard.
REFERENCE: ITP POWERPLANT, Chapter 8, Page 6 (Figure 1A-4)
               AC65-12A, Page 296 (Figure 6-11)

4393. ANSWER #2

The purpose of an oil pressure relief valve is to maintain pressure at a specified level. On most engines, the desired pressure is reached before the engine reaches cruise RPM. This means that at cruise the relief valve is off its seat, bypassing some of the oil back to the inlet of the oil pump.
REFERENCE: ITP POWERPLANT, Chapter 8, Pages 6 and 7

4394. ANSWER #4

In gas turbine engines, fuel and ram air are both used to cool the oil. The fuel-cooled oil cooler acts as a fuel/oil heat exchanger in that the fuel cools the oil and the oil heats the fuel. The air-cooled oil cooler normally is installed at the forward end of the engine and is similar to those used on recip engines.
REFERENCE: ITP POWERPLANT, Chapter 8, Page 26
               AC65-12A, Page 308

4395. ANSWER #4

Most oil temperature bulbs in recip engines are mounted in the pressure oil screen housing. It is located so that it measures the oil temperature before it enters the engine's hot section.
REFERENCE: ITP POWERPLANT, Chapter 8, Page 15
               AC65-12A, Page 299

4396. ANSWER #1

If an oil pump in a newly installed engine is turning, with an adequate supply of oil, a possible reason for it not building up any pressure is excessive pump side clearances. Excessive side clearances will make it difficult for the pump to pick up a prime, and in a newly installed engine the pump must pick up a prime if it is going to build pressure.
REFERENCE: AIRCRAFT POWERPLANTS, Page 351

4397. ANSWER #3

The flow control valve determines which of the two possible paths the oil will take through a cooler. When the oil is cold, a bellows within the flow control valve contracts and lifts a valve from its seat. In this open position, the oil entering the cooler has a choice of two outlets and two paths. Following the path of least resistance, the oil flows around the jacket and out past the thermostatic valve to the tank. If the control valve sticks in the open position, this is the path oil will follow.
REFERENCE: ITP POWERPLANT, Chapter 8, Page 26
               AC65-12A, Page 291

4398. ANSWER #3

In addition to the main oil filters in an oil system, there are also secondary filters located throughout the system for various purposes. For instance, there may be a finger screen filter, which is sometimes used for straining scavenged oil. Also, there are fine-mesh screens, called "last chance" filters, for straining the oil just before it passes from the spray nozzles onto the bearing surfaces.
REFERENCE: ITP POWERPLANT, Chapter 8, Page 26
               AC65-12A, Page 306

4399. ANSWER #4

Thermostatic bypass valves are included in oil systems using an oil cooler. Their purpose is to always maintain proper oil temperature by varying the proportion of the total oil flow passing through the oil cooler.
REFERENCE: ITP POWERPLANT, Chapter 8, Page 26
               AC65-12A, Page 308

4400. ANSWER #3

Recip engine oil tanks are fitted with vent lines to ensure proper tank ventilation in all attitudes of flight. These lines are usually connected to the engine crankcase to prevent the loss of oil through the vents. This indirectly vents the tanks to the atmosphere through the crankcase breather.
REFERENCE: AC65-12A, Pages 286 and 287

4401. ANSWER #4
The exhaust turbine bearing is the most critical lubricating point in a gas turbine engine because of the high temperature normally present. In some engines, air cooling is used in addition to oil cooling for the bearing which supports the turbine.
REFERENCE: ITP POWERPLANT, Chapter 7, Page 30
            AC65-12A, Page 302

4402. ANSWER #4
As identified in question #4401, the turbine bearing is the most critical lubricating point in a gas turbine engine. In most turbine engines, the quantity of oil supplied to the turbine bearing or bearings is greater than to any of the other engine bearings. This is necessary because of the tremendous amount of heat which must be carried away to keep the bearing cool.
REFERENCE: ITP POWERPLANT, Chapter 7, Page 30
            AC65-12A, Page 302

4403. ANSWER #4
The fuel/oil heat exchanger is designed to cool the hot oil and to preheat the fuel for combustion. Fuel flowing to the engine must pass through the heat exchanger; however, there is a thermostatic valve which controls the oil flow, and the oil may bypass the cooler if no cooling is needed.
REFERENCE: ITP POWERPLANT, Chapter 8, Page 26
            AC65-12A, Page 308

4404. ANSWER #2
Oil tank filler openings on turbine engines must be marked with the word "oil" and the capacity of the tank.
REFERENCE: ITP POWERPLANT, Chapter 8, Page 16
            FAR 33.71 (c) (5)

4405. ANSWER #1
Turbine engine oil tanks, after being repaired, should be pressure checked to the same pressure requirements needed for certification. These requirements are given in FAR 33, and call for a pressure of 5 PSI.
REFERENCE: FAR 33.71 (c) (9)

4406. ANSWER #1
Pressurized oil distributed to a turbine engine's main bearings is sprayed on the bearings through fixed orifice nozzles, thus providing a relatively constant oil flow at all engine operating speeds.
REFERENCE: AC65-12A, Page 309

4407. ANSWER #3
The AC65-12A supplies answers #2 and #3 as being correct. Answer #3, the oil tank serving as both a reservoir and cooler, is a feature of many of the early turbine engines which had wet-sump oil systems. Answer #2 is not a good answer because dry-sump engines use air to cool the turbines to a much greater degree than the wet-sump engines ever did.
REFERENCE: AC65-12A, Page 308

4408. ANSWER #3
When the oil pressure in a turbine engine lubrication system has reached the desired value, the relief valve unseats and bypasses just enough oil back to the inlet of the pump to maintain that pressure. If the relief valve were to stick in the open position, it would bypass oil continuously and take away necessary lubrication from the engine.
REFERENCE: ITP POWERPLANT, Chapter 8, Page 24
            AC65-12A, Pages 306 and 307

4409. ANSWER #2
An oil to fuel heat exchanger in a turbine engine acts to cool the oil and preheat the fuel. Of these two functions, the primary purpose is to cool the oil. This is known because the fuel flows through the heat exchanger all the time. The oil only flows through when it needs to be cooled.
REFERENCE: ITP POWERPLANT, Chapter 8, Page 26
            AC65-12A, Page 308

4410. ANSWER #1
In an aircraft engine's lubrication system, the oil pressure relief valve is adjusted to maintain the desired system pressure. The pressure is maintained by bypassing oil back to the inlet side of the pump when the pressure tries to climb too high.
REFERENCE: ITP POWERPLANT, Chapter 8, Page 24
            AC65-12A, Pages 300, 306 and 307

4411. ANSWER #2

An oil pressure relief valve limits oil pressure to the value specified by the engine manufacturer. The oil pressure in an engine must be high enough to ensure adequate lubrication of the engine and accessories at high speeds and power. On the other hand, the pressure must not be too high, since leakage and damage to the oil system may result.
REFERENCE: AC65-12A, Page 300

4412. ANSWER #4

The breather pressurizing system of a turbine engine ensures a proper spray pattern from the main bearing oil jets and furnishes a pressure head to the scavenge system. If the pressure in the bearing housings was allowed to drop off too much, as atmospheric pressure dropped off, the flow of oil from the oil jets would change. To maintain a relatively constant flow rate and provide a head of pressure for the scavenge system, the breather pressurizing system maintains a pressure higher than ambient as the aircraft climbs to altitude.
REFERENCE: ITP POWERPLANT, Chapter 8, Page 27
              AC65-12A, Page 312

4413. ANSWER #4

Directing bleed air to the bearings on turbine engines, for the purpose of cooling, relieves the oil system of some of its job of cooling. It is possible that this might make an oil cooler unnecessary, but that would depend on the particular engine. There are a few wet-sump engines in use today, and some do use oil coolers.
REFERENCE: AC65-12A, Page 302

4414. ANSWER #1

The purpose of a bypass valve in an oil cooler is to direct the oil around the cooler and back to the oil tank when the oil is cold or when the core is blocked with thick, congealed oil.
REFERENCE: ITP POWERPLANT, Chapter 8, Page 11
              AC65-12A, Pages 290 and 291

4415. ANSWER #4

In an aircraft engine's lubrication system, the oil pressure relief valve is adjusted to maintain the desired system pressure. The pressure is maintained by bypassing oil back to the inlet side of the pump when the pressure tries to climb too high.
REFERENCE: ITP POWERPLANT, Chapter 8, Pages 9, 10 and 24
              AC65-12A, Pages 300, 306 and 307

4416. ANSWER #2

As oil circulates through a reciprocating engine, it absorbs heat from the parts. The pistons and cylinders are especially dependent on the oil for cooling because they are closest to the temperatures of combustion.
REFERENCE: AC65-12A, Page 285

4417. ANSWER #1

Gears in the accessory section of an engine are normally lubricated by splashed or sprayed oil.
REFERENCE: ITP POWERPLANT, Chapter 8, Page 7

4418. ANSWER #3

In a dry-sump oil sytem for a recip engine, there is often a check valve installed in the oil filter. This check valve is closed by a light spring loading of 1 to 3 pounds when the engine is not operating, to prevent gravity-fed oil from entering the engine. This could be a serious problem in a radial engine if the oil got into the lower cylinders and caused a liquid lock. Turbine engines also make use of check valves to prevent oil from leaking into the engine.
REFERENCE: ITP POWERPLANT, Chapter 8, Page 5
              AC65-12A, Pages 288 and 307

4419. ANSWER #1

Although the mechanical condition of an engine will affect its oil consumption, the mechanical efficiency does not really have any effect. Mechanical efficiency would deal with things like energy losses in a gearbox, or how efficiently the propeller converts power into thrust. These factors have little if any effect on oil consumption. Engine temperature, RPM, and lubricant characteristics certainly do.
REFERENCE: AIRCRAFT POWERPLANTS, Page 353

4420. ANSWER #3

Oil control rings regulate the thickness of the oil film on the cylinder wall. To allow the surplus oil to return to the crankcase, holes are drilled in the piston ring grooves or in the lands next to these grooves.
REFERENCE: ITP POWERPLANT, Chapter 1, Page 33
              AC65-12A, Page 17

4421. ANSWER #3
The fuel line for the oil dilution vlave should never be located between the pressure pump and the engine pressure system. The connection should not be in the pressure system because the oil under pressure could end up entering the fuel system instead of the fuel entering the oil system. The connection for dry-sump systems is usually made at the oil drain Y-valve.
REFERENCE: ITP POWERPLANT, Chapter 8, Page 11
                       AC65-12A, Page 287

4422. ANSWER #4
Oil dilution systems operate by thinning the oil with fuel just before the engine is shut down.
REFERENCE: ITP POWERPLANT, Chapter 8, Page 11
                       AC65-12A, Page 287

4423. ANSWER #4
The initial adjustment on the oil pressure relief valve for a newly overhauled engine is made in the overhaul shop. The adjustment is fine tuned when the engine is installed in a test cell and run. The initial adjustment can't wait until the engine is run because of the possiblity of starving the engine for oil.
REFERENCE: AC65-12A, Page 300

4424. ANSWER #1
In dry-sump lubricating systems the oil temperature bulb may be anywhere in the oil inlet line between the supply tank and the engine.
REFERENCE: ITP POWERPLANT, Chapter 8, Page 15
                       AC65-12A, Page 290

4425. ANSWER #1
In most recip engines, the cylinder walls receive oil sprayed from the crankshaft and also from the crankpin bearings. Some of the oil coming from the crankshaft is also being splashed onto the cylinder walls.
REFERENCE: ITP POWERPLANT, Chapter 8, Page 7
                       AC65-12A, Page 294

4426. ANSWER #4
Full-flow oil filters used on aircraft engines are equipped with a bypass valve which allows unfiltered oil to bypass the filter and enter the engine when the oil filter is clogged or during a cold weather start.
REFERENCE: ITP POWERPLANT, Chapter 8, Page 7
                       AC65-12A, Page 288

4427. ANSWER #4
Some of the cylinders in a radial engine and all the cylinders in an inverted engine are located at the bottom of the engine. It is necessary, therefore, to incorporate features to prevent these cylinders from being flooded with oil. This is accomplished by means of long skirts on the cylinders and an effective scavenging system.
REFERENCE: AIRCRAFT POWERPLANTS, Page 182

4428. ANSWER #3
The best method of lubricating engine roller bearings and gear systems is to operate them in a housing to which oil under pressure is supplied. The bearings should have pressure oil supplied to them directly and the gears should be lubricated by a splash system.
REFERENCE: ITP POWERPLANT, Chapter 8, Page 7

4429. ANSWER #2
Oil in service, especially in recip engines, is constantly exposed to many harmful substances that reduce its ability to protect moving parts. The main contaminants are gasoline, moisture, acids, dirt, carbon, and metallic particles. Because of the accumulation of these harmful substances, common practice is to drain the entire lubrication system at regular intervals and refill with new oil.
REFERENCE: AC65-12A, Pages 300 and 301

4430. ANSWER #3
The cuno-type oil flter has a cartridge made of disks and spacers. A cleaner blade fits between each pair of disks. Oil from the pump enters the cartridge well that surrounds the cartridge and passes through the spaces between the closely spaced disks of the cartridge, then through the hollow center and on to the engine. The amount of space between the disks determines the size of particle that the filter will hold and not allow to pass out to the engine.
REFERENCE: AC65-12A, Page 289

4431. ANSWER #1

Some oil tanks have a built-in hopper, or temperature accelerating well, that extends from the oil return fitting on top of the oil tank to the outlet fitting in the sump in the bottom of the tank. By separating the circulating oil from the surrounding oil in the tank, less oil is circulated. This hastens the warming of the oil when the engine is started.
REFERENCE: AC65-12A, Page 287

4432. ANSWER #1

The flow control valve determines which of the two possible paths the oil will take through a cooler. When the oil is cold, the valve directs oil around the cooler and back to the tank (or out to the engine if it is a hot tank system). When the oil is hot, the valve directs oil through the cooler.
REFERENCE: ITP POWERPLANT, Chapter 8, Page 11
                         AC65-12A, Page 291

4433. ANSWER #1

Some crankshafts are manufactured with hollow crankpins that serve as sludge removers. The sludge chambers may be formed by means of spool-shaped tubes pressed into the hollow crankpins or by plugs pressed into each end of the crankpin. The chambers are cleaned at overhaul.
REFERENCE: ITP POWERPLANT, Chapter 1, Page 35 (Figure 4A-17)
                         AC65-12A, Page 428

4434. ANSWER #3

Oil tanks are equipped with vent lines to ensure proper tank ventilation in all attitudes of flight. These lines are usually connected to the engine crankcase to prevent the loss of oil through the vents. This indirectly vents the tank to the atmosphere through the crackcase breather.
REFERENCE: AC65-12A, Pages 286 and 287

4435. ANSWER #4

Oil control rings regulate the thickness of the oil film on the cylinder wall. To allow the surplus oil to return to the crankcase, holes are drilled in the piston ring grooves or in the lands next to these grooves.
REFERENCE: ITP POWERPLANT, Chapter 1, Page 33
                         AC65-12A, Page 17

4436. ANSWER #3

Although some of the early gas turbine engines built made use of wet-sump oil systems, they are seldom used anymore. The dry-sump system is the most popular in modern gas turbine engines. Because gas turbine engines use synthetic oil, which has flow characteristics far superior to the mineral base oil used in recip engines, oil dilution is not needed in cold weather.
REFERENCE: ITP POWERPLANT, Chapter 8, Page 16
                         AC65-12A, Pages 302 and 304

4437. ANSWER #1

When the circulating oil has performed its function of lubricating and cooling the moving parts of the engine, it drains into the sumps in the lowest parts of the engine. Oil collected in these sumps is picked up by the scavenge oil pumps as quickly as it accumulates. These pumps have a greater capacity than the pressure pump, because the oil they move is heated and therefore occupies a greater volume.
REFERENCE: ITP POWERPLANT, Chapter 8, Page 6
                         AC65-12A, Page 290

4438. ANSWER #3

The flow control valve on an oil cooler (also known as a bypass valve or thermostatic valve) deter-mines whether the oil flows through the cooler or flows around it. The valve is open when the oil is cold and closes as the oil heats up. The valve is open the greatest amount when the oil is colder than normal.
REFERENCE: ITP POWERPLANT, Chapter 8, Page 11 (Figure 1A-11)
                         AC65-12A, Page 291

4439. ANSWER #2

The Federal Aviation Regulations require that the word "oil" and the oil tank capacity be marked on or near the filler cover of all turbine engine oil tanks.
REFERENCE: ITP POWERPLANT, Chapter 8, Page 16
                         FAR 33.71 (c) (5)

4440. ANSWER #1

The variation in the speed of an oil pump from idling to full-throttle operation and the fluctuation of oil viscosity because of temperature changes are compensated for by the tension on the relief valve spring. The pump is designed to create a greater pressure than probably will ever be required, to compensate for wear of the bearings or thinning out of the oil.
REFERENCE: AC65-12A, Page 294

4441. ANSWER #3
The overhead valve assemblies of opposed engines used in helicopters and in airplanes are lubricated by the pressure system. Oil under pressure flows through the hydraulic tappet body and through hollow pushrods to the rocker arm, where it lubricates the rocker arm bearing and the valve stem.
REFERENCE: ITP POWERPLANT, Chapter 8, Pages 6 and 7
               AC65-12A, Pages 294 and 295

4442. ANSWER #4
Aircraft engine oil filters are equipped with a bypass valve, to allow oil to flow around the filter in the event it clogs or the oil is too congealed to flow through it. Providing the oil pressure relief valve is downstream of the filter, the bypass valve will allow the oil flow to the system to be normal.
REFERENCE: ITP POWERPLANT, Chapter 8, Page 7
               AC65-12A, Page 288

4443. ANSWER #3
The dry-sump lubrication system of a typical turbine engine consists of the pressure, scavenge, and breather subsystems. The pressure system supplies oil to the main engine bearings and to the accessory drives. The scavenge system returns the oil to the engine oil tank. The breather system connects the individual bearing compartments and the oil tank and vents them to atmosphere, possibly through a breather pressurizing valve.
REFERENCE: ITP POWERPLANT, Chapter 8, Page 26
               AC65-12A, Page 308
               AC65-12A, Page 307

4444. ANSWER #3
Last chance filters in turbine engines are generally located right at the oil jets. Because these jets are inside the bearing housings, the last chance filters can be cleaned only at overhaul.
REFERENCE: ITP POWERPLANT, Chapter 8, Page 26
               AC65-12A, Page 307

4445. ANSWER #2
The piston pins on most recip engines are lubricated by oil which is sprayed or thrown from the master or connecting rod.
REFERENCE: ITP POWERPLANT, Chapter 8, Pages 8 and 9 (Figure 1A-6)

4446. ANSWER #3
Oil tanks are equipped with vent lines to ensure proper tank ventilation in all attitudes of flight. These lines are usually connected to the engine crankcase to prevent the loss of oil through the vents. This indirectly vents the tank to the atmosphere through the crankcase breather.
REFERENCE: AC65-12A, Pages 286 and 287

4447. ANSWER #3
An oil pressure relief valve in an engine oil system is located downstream of the pressure pump and upstream of the internal oil system.
REFERENCE: ITP POWERPLANT, Chapter 8, Page 8 (Figure 1A-6)
               AC65-12A, Page 288 (Figure 6-4)

4448. ANSWER #3
When the circulating oil has performed its function of lubricating and cooling the moving parts of the engine, it drains into the sumps in the lowest parts of the engine. Oil collected in these sumps is picked up by the scavenge oil pumps as quickly as it accumulates. These pumps have a greater capacity than the pressure pump, because the oil they move is heated and therefoe occupies a greater volume.
REFERENCE: ITP POWERPLANT, Chapter 8, Page 6
               AC65-12A, Page 290

4449. ANSWER #1
The temperature control valve (also called a flow control valve or thermostatic valve) is located at the inlet to the cooler. It is this valve which controls the flow of oil through or around the cooler.
REFERENCE: ITP POWERPLANT, Chapter 8, Page 11
               AC65-12A, Page 291

4450. ANSWER #3
The flow control valve on an oil cooler (also known as a bypass valve or thermostatic valve) determines whether the oil flows through the cooler or flows around it. The valve is open when the oil is cold and closes as the oil heats up. The valve is open the greatest amount when the oil is colder than normal.
REFERENCE: ITP POWERPLANT, Chapter 8, Page 11 (Figure 1A-11)
               AC65-12A, Page 291

4451. ANSWER #1
In most turbine engine oil tanks, a pressure build-up is desired to assure a positive flow of oil to the oil pump inlet. This pressure build-up is accomplished by running the tank overboard vent line through an adjustable relief valve to maintain a positive pressure of approximately 3 to 6 psig.
REFERENCE: ITP POWERPLANT, Chapter 8, Page 19

4452. ANSWER #4
An oil pressure relief valve in a recip engine bypasses oil back to the inlet side of the pressure pump. Only enough oil is bypassed to maintain the desired oil pressure in the system.
REFERENCE: ITP POWERPLANT, Chapter 8, Page 9
            AC65-12A, Page 288 (Figure 6-4)

4453. ANSWER #2
The thermostatic control valve in an oil cooler (also known as a thermostatic bypass valve or surge protection valve) is designed to let oil bypass the oil cooler if the oil is congealed. **NOTE:** Among the questions dealing with oil systems, it is interesting to note that this valve has been called by the following names:
1. Engine oil temperature regulator, question #3391.
2. Thermostatic-type oil temperature control valve, question #3397.
3. Automatic bypass valve, question #4438.
4. Temperature control valve, question #4449.
5. Oil cooler flow control valve, question #4450.
6. Thermostatic control valve, question #4453.
For all practical purposes, these valves, by any name, do the same thing.
REFERENCE: ITP POWERPLANT, Chapter 8, Page 11
            AC65-12A, Page 291

4454. ANSWER #3
The primary source of oil contamination in a recip engine is combustion by-products escaping past the piston rings, known as blow-by, and the carbonizing of the oil which overheats because it is trapped in the pores of the cylinder walls.
REFERENCE: AC65-12A, Page 300

4455. ANSWER #3
When oil pressure in an engine tries to climb too high, the relief valve unseats and bypasses oil back to the inlet of the pressure pump. If the relief valve were to stick open, oil would be bypassed even when the pressure was not too high. This would cause a reduced amount of oil to flow to the engine, which would cause low oil pressure. The reduction in oil flow would also cause high oil temperature because there would be less cooling oil available.
REFERENCE: ITP POWERPLANT, Chapter 1, Page 77
            AC65-12A, Page 301

4456. ANSWER #4
In an aircraft engine's oil system, main oil filters are located right after the pressure pump.
REFERENCE: ITP POWERPLANT, Chapter 8, Page 7
            AC65-12A, Page 288

4457. ANSWER #3
In dry-sump lubrication systems, both recip and turbine, check valves are installed somewhere between the oil tank and the engine proper. These valves are set at 2 to 5 psi, and their purpose is to prevent oil from draining into the engine when it is not operating. Without the check valve, the oil in the tank would flow into the engine and flood the gearboxes and accessory cases.
REFERENCE: ITP POWERPLANT, Chapter 8, Page 5
            AC65-12A, Pages 288 and 307

4458. ANSWER #1
Oil coolers are found more often on turbine engine dry-sump systems than on wet-sump systems. Wet-sump sytems on turbine engines, of course, are few and far between.
REFERENCE: AC65-12A, Page 302

4459. ANSWER #2
According to the Federal Aviation Regulations, an oil tank must have an expansion space of 10% or 0.5 gallon, whichever is greater. In this case they are the same.
REFERENCE: ITP POWERPLANT, Chapter 8, Page 16
            FAR 23.1013 (b) (1)

4460. ANSWER #4
The flapper valves in a turbine engine oil tank are located in the baffle between the top and bottom of the tank. The flapper valves are normally open, allowing oil to flow between the two sections of the tank. The valves only close when the oil in the bottom of the tank tends to rush to the top of the tank during deceleration.
REFERENCE: AC65-12A, Page 304

4461. ANSWER #2
All oil tanks are provided with expansion space. This allows for expansion of the oil after heat is absorbed from the bearings and gears and after the oil foams as a result of circulating through the system.
REFERENCE: AC65-12A, Page 304

4462. ANSWER #4
Some turbine engine oil tanks incorporate a de-aerator tray for separating air from the oil returned to the top of the tank by the scavenge system. Usually these de-aerators are the "can" type.
REFERENCE: ITP POWERPLANT, Chapter 8, Page 20
                         AC65-12A, Page 304

4463. ANSWER #4
All bearings require lubrication. Plain bearings (friction bearings) must have oil supplied to them under pressure to prevent metal to metal contact. Ball, roller, and tapered roller bearings can survive with oil being sprayed on them, as they do in the turbine engine.
REFERENCE: AC65-12A, Pages 293 and 310

4464. ANSWER #2
During the overhaul of a magneto, the magnet should be handled carefully and should have a soft iron keeper of the proper shape placed over the poles to prevent loss of magnetism.
REFERENCE: AIRCRAFT POWERPLANTS, Page 260

4465. ANSWER #1
There are two common methods of checking the strength of the magnet for a magneto. If the magnet is removed during overhaul, it can be checked with a magnetometer. If the magneto is assembled, the strength can be checked by rotating the magneto at a specified speed with the points held open and checking the output of the primary with an a.c. ammeter.
REFERENCE: AIRCRAFT POWERPLANTS, Page 260

4466. ANSWER #4
The dwell angle of a magneto is the number of degrees of rotation that the breaker points are closed. When dealing with magnetos, it is the E-gap angle which is considered and not the dwell angle. Dwell angle is only important when working on a battery ignition system.
REFERENCE: AIRCRAFT POWERPLANTS, Page 226 (Figure 10-10)

4467. ANSWER #2
The number of degrees between the neutral position and the position where the contact points open is called th E-gap angle. The manufacturer of the magneto determines for each model how many degrees beyond the neutral position a pole of the rotor magnet should be in order to obtain the strongest spark.
REFERENCE: ITP POWERPLANT, Chapter 3, Page 10 (Figure 2-5)
                         AIRCRAFT POWERPLANTS, Page 226

4468. ANSWER #2
When the magnet is in a position where the magnetically opposite poles are perfectly aligned with the pole shoes, the number of magnetic lines of force through the coil core is maximum. This is called the "full register" position and produces a maximum number of magnetic lines of force, known as the flux flow.
REFERENCE: ITP POWERPLANT, Chapter 3, Page 7
                         AC65-12A, Page 178

4469. ANSWER #3
The internal timing of a magneto can be most easily set when the magneto is being assembled after overhaul or maintenance.
REFERENCE: ITP POWERPLANT, Chapter 3, Pages 17 and 18
                         AC65-12A, Pages 201 and 202

4470. ANSWER #3
The ignition harness of a recip engine contains an insulated wire for each cylinder that the magneto serves in the engine. The harness serves a dual purpose. It supports the wires and protects them from damage by engine heat, vibration, or weather. It also serves as a conductor for stray magnetic fields that surround the wires as they momentarily carry high-voltage current.
REFERENCE: AC65-12A, Page 184

4471. ANSWER #2
A magneto is internally timed so the breaker points begin to open at the E-gap angle. The magneto is then timed to the engine so the opening of the points produces a spark in the cylinder at the proper time. If the point gap is arbitrarily increased, the spark will come early, and it will be of a decreased intensity because the magneto is not at E-gap.
REFERENCE: ITP POWERPLANT, Chapter 3, Pages 24 and 25
                         AC65-12A, Pages 180, 202 and 203

4472. ANSWER #1

Magnetos are sometimes equipped with a safety gap to provide a return ground when the external secondary circuit is open. It is connected in series with the secondary circuit and hence protects it against shorts or open circuits. The safety gap protects against damage from excessively high voltage in case the secondary circuit is accidentally broken and the spark cannot jump between the electrodes of the spark plug.
REFERENCE: AIRCRAFT POWERPLANTS, Page 229

4473. ANSWER #3

When internally timing a magneto, the magnets are set in the E-gap position and the breaker points are just starting to open. It might vary from one magneto to another, but most have a set of aligment marks which should line up when the magneto is properly timed internally.
REFERENCE: ITP POWERPLANT, Chapter 3, Page 22
             AC65-12A, Page 201

4474. ANSWER #4

When internally timing a magneto , the breaker points should just begin to open when the magnet is a few degrees past the neutral position. The neutral position is 45 degrees from the full register position, meaning 45 degrees from the time when the poles are lined up with the pole shoes.
REFERENCE: ITP POWERPLANT, Chapter 3, Page 9
             AC65-12A, Page 201

4475. ANSWER #2

The primary electrical circuit in a magneto consists of a set of breaker contact points, a condenser, and an insulated coil. The condenser is wired in parallel with the breaker points. Its purpose is to prevent arcing of the points when the circuit is opened, and it hastens the collapse of the magnetic field about the primary coil.
REFERENCE: ITP POWERPLANT, Chapter 3, Page 10
             AC65-12A, Page 179

4476. ANSWER #3

With the magneto in E-gap position, a very high rate of flux change can be obtained by opening the primary breaker points. Opening the breaker points stops the flow of current in the primary circuit, and allows the magnetic rotor to quickly reverse the field through the coil core. This sudden flux reversal produces a high rate of flux change in the core, which cuts across the secondary coil of the magneto and induces a pulse of high-voltage current in the secondary. This is the charge that fires the spark plug.
REFERENCE: ITP POWERPLANT, Chapter 3, Pages 4, 9 and 10
             AC65-12A, Page 179

4477. ANSWER #4

The purpose of an impulse coupling is to spin the magneto faster than normal during starting and to retard the spark. While the engine is running, the impulse coupling does not serve any purpose. Some magnetos will operate normally with a broken impulse coupling spring, except during starting. Other magnetos will operate with a retarded spark if the spring breaks. The possible answers in this question only mention the normal operation.
REFERENCE: ITP POWERPLANT, Chapter 3, Pages 16 and 17
             AC65-12A, Page 193

4478. ANSWER #1

The two north poles of a four-pole rotating magnet are located 180 degrees apart.
REFERENCE: AC65-12A, Page 178 (Figure 4-2)

4479. ANSWER #1

Magneto pole shoes and their extensions are generally made of soft-iron laminations cast in the magneto housing. The coil core is also made of soft-iron laminations.
REFERENCE: AIRCRAFT POWERPLANTS, Page 223

4480. ANSWER #1

When routing long lengths of ignition cable through a rigid ignition harness, the cable should be dusted with soapstone to reduce the friction and therefore the possibility of damage.
REFERENCE: AIRCRAFT POWERPLANTS, Page 228
             AIRCRAFT PROPULSION POWERPLANTS, Page 302

4481. ANSWER #3

The magnetic circuit of a magneto consists of a permanent multipole rotating magnet, a soft-iron core, pole shoes, and the pole shoe extensions.
REFERENCE: ITP POWERPLANT, Chapter 3, Page 7
             AC65-12A, Page 177

4482. ANSWER #2
The purpose of the condenser in the primary electrical circuit is to prevent arcing at the points when the circuit is opened, and to hasten the collapse of the magnetic field about the primary coil.
REFERENCE: ITP POWERPLANT, Chapter 3, Pages 3 and 10
                    AC65-12A, Page 179

4483. ANSWER #2
With the magneto in E-gap position, a very high rate of flux change can be obtained by opening the primary breaker points. Opening the breaker points stops the flow of current in the primary circuit, and allows the magnetic rotor to quickly reverse the field through the coil core. This sudden flux reversal produces a high rate of flux change in the core, which cuts across the secondary coil of the magneto and induces the pulse of high-voltage current in the secondary. This is the charge that fires the spark plug.
REFERENCE: ITP POWERPLANT, Chapter 3, Pages 4, 9 and 10
                    AC65-12A, Page 179

4484. ANSWER #3
The switch in a battery ignition system operates a bit differently than the switch in a magneto ignition system. When the battery ignition switch is on, it supplies current to the primary of the coil. When it is off, it opens the circuit to the primary of the coil. When a magneto ignition switch is off, it grounds the primary circuit.
REFERENCE: ITP POWERPLANT, Chapter 3, Page 2
                    AC65-12A, Page 177 (Figure 4-1)

4485. ANSWER #2
To produce sparks, the rotating magnet in a magneto must be turned at or above a specified number of revolutions per minute, at which speed the rate of change in flux linkages is sufficiently high to induce the required primary current and the resultant high-tension output. The higher the speed of the magneto, the greater the voltage induced in the primary.
REFERENCE: AIRCRAFT POWERPLANTS, Page 228

4486. ANSWER #2
The breaker points in a magneto begin to open a few degrees past the neutral position, called E-gap angle. At this time, the flux flow is zero. The induced current in the primary circuit at this time is so strong, there can be no flux flow in the magnetic circuit. The moment the points open, all the fields collapse and flux flow starts again in the magnetic circuit.
REFERENCE: ITP POWERPLANT, Chapter 3, Pages 7 and 8
                    AC65-12A, Page 178

4487. ANSWER #4
The firing order of a 9 cylinder radial engine is 1, 3, 5, 7, 9, 2, 4, 6, 8. The sixth cylinder to fire is #2, so this cylinder would be fired by the number 6 distributor block electrode.
REFERENCE: ITP POWERPLANT, Chapter 3, Page 11
                    AC65-12A, Page 183

4488. ANSWER #2
Magnetos are sometimes equipped with a safety gap to provide a return ground when the external secondary circuit is open. It is connected in series with the secondary circuit and hence protects it against shorts or open circuits. The safety gap protects against damage from excessively high voltage in case the secondary circuit is accidentally broken and the spark cannot jump between the electrodes of the spark plug.
REFERENCE: AIRCRAFT POWERPLANTS, Page 229

4489. ANSWER #3
The firing order of a 7 cylinder radial engine is 1, 3, 5, 7, 2, 4, 6. The third cylinder to fire is number 5, so the number 3 distributor wire would fire this cylinder.
REFERENCE: ITP POWERPLANT, Chapter 1, Page 24
                    AC65-12A, Pages 19 and 182

4490. ANSWER #3
A defective primary condenser will be most noticed by a burned and pitted set of breaker points. A defective condenser will not be able to absorb the self-induced current flowing in the primary circuit, so the current will arc across the breaker points and cause the burning and pitting. A good set of breaker points will have a dull gray or sandblasted appearance, something that could be described as a frosty look. In reference to this question, the AC65-12A is misleading and in error.
REFERENCE: ITP POWERPLANT, Chapter 3, Page 10
                    AIRCRAFT POWERPLANTS, Page 227

4491. ANSWER #1

In a low-tension ignition system, the secondary coils are located at the cylinders. There is a separate secondary coil for each of the spark plugs. Since an 18 cylinder engine will have 36 spark plugs, there must be 36 secondary coils.
REFERENCE: AC65-12A, Page 186

4492. ANSWER #3

The ignition switch in a magneto circuit is wired parallel to the breaker points. From the ignition switch, there are two paths which can be followed; one is through the breaker points and the other is through the primary of the coil.
REFERENCE: ITP POWERPLANT, Chapter 3, Page 3
             AC65-12A, Pages 184 and 185 (Figure 4-12)

4493. ANSWER #3

With the magneto in E-gap position, a very high rate of flux change can be obtained by opening the primary breaker points. Opening the breaker points stops the flow of current in the primary circuit, and allows the magnetic rotor to quickly reverse the field through the coil core. This sudden flux reversal produces a high rate of flux change in the core, which cuts across the secondary coil of the magneto and induces a pulse of high-voltage current in the secondary. This is the charge that fires the spark plug.
REFERENCE: ITP POWERPLANT, Chapter 3, Pages 4, 9 and 10
             AC65-12A, Page 179

4494. ANSWER #4

The shielding used on spark plug and ignition wires serves to prevent or reduce interference with radio reception. Without this shielding, radio communication would become virtually impossible.
REFERENCE: AC65-12A, Page 184

4495. ANSWER #3

The impulse coupling is a unit which, at the time of spark production, gives one of the magnetos attached to the engine a brief acceleration to help it produce a hot spark for starting. In addition to helping produce a hot spark, it also retards the spark a few degrees to aid in starting.
REFERENCE: ITP POWERPLANT, Chapter 3, Pages 16 and 17
             AC65-12A, Page 193

4496. ANSWER #3

Dual-ignition spark plugs may be set to fire at the same instant (synchronized) or at slightly different intervals (staggered). When staggered ignition is used, each of the two sparks occur at a different time. The spark plug in the exhaust side of the cylinder always fires first because the slower rate of burning of the expanded and diluted fuel-air mixture at this point in the cylinder makes it desirable to have an advance in the ignition timing.
REFERENCE: AIRCRAFT POWERPLANTS, Page 221

4497. ANSWER #4

Dual-ignition spark plugs may be set to fire at the same instant (synchronized) or at slightly different intervals (staggered). When staggered ignition is used, each of the two sparks occur at a different time. The spark plug in the exhaust side of the cylinder always fires first because the slower rate of burning of the expanded and diluted fuel-air mixture at this point in the cylinder makes it desirable to have advance in the ignition timing.
REFERENCE: AIRCRAFT POWERPLANTS, Page 221

4498. ANSWER #3

Magnetos cannot be hermetically sealed to prevent moisture from entering a unit because the magneto is subject to pressure and temperature changes in altitude. Thus, adequate drains and proper ventilation reduce the tendency of flashover and carbon tracking. Good magneto circulation also ensures that corrosive gases produced by normal arcing across the distributor air gap are carried away.
REFERENCE: AC65-12A, Page 183

4499. ANSWER #2

This question does not have a good answer, but the answer that was probably intended to be correct is #2. If there is an open in the primary lead connecting the ignition switch to the primary circuit of the magneto, the engine will not cease firing when the switch is turned off. The high tension lead in an ignition system is the lead which fires the spark plug. This would have nothing to do with an engine failing to quit running.
REFERENCE: ITP POWERPLANT, Chapter 3, Page 24
             AC65-12A, Page 211

4500. ANSWER #2

Internal timing of a magneto means ensuring that the rotor magnet is the proper number of degrees past the neutral position, called the E-gap angle, and that the breaker points are just starting to open. Alignment of the marks provided should provide this condition.
REFERENCE: ITP POWERPLANT, Chapter 3, Page 22
             AC65-12A, Page 201

4501. ANSWER #1
When using a timing light to check a magneto in a complete ignition system installed on the aircraft, the master ignition switch for the aircraft must be turned on and the ignition switch for the engine turned to "both". Otherwise, the lights will not indicate breaker point opening.
REFERENCE: AC65-12A, Page 203

4502. ANSWER #4
Electronically, the low-tension ignition system is different from the high-tension system. In the low-tension system, low voltage is generated in the magneto and flows to the primary winding of a transformer coil located near the spark plug. In the high-tension system, high voltage is generated in the magneto and travels through high-tension leads to fire the spark plug.
REFERENCE: ITP POWERPLANT, Chapter 3, Page 2
           AC65-12A, Page 186

4503. ANSWER #1
Several different types of test devices are used for determining the serviceablility of a high-tension ignition harness. One common type of tester is capable of applying a direct current in any desired voltage from 0 to 15,000 volts with a 110 volt, 60 cycle input. The current leakage between ignition cable and manifold is measured on two scales of a microammeter that is graduated to read from 0 to 100 microamps and 0 to 1,000 microamps.
REFERENCE: AC65-12A, Page 226

3504. ANSWER #2
In the primary circuit of the magneto, it is the opening of the breaker which causes the collapse of the primary field. The collapse of the primary is what induces the high voltage in the secondary which fires the spark plug. If the breaker points stick in the open position, the primary will continue to produce current, but the magneto will not produce current to fire the plug because there will be no induced voltage in the secondary.
REFERENCE: ITP POWERPLANT, Chapter 3, Pages 9 and 10
           AC65-12A, Pages 180 and 220

4505. ANSWER #4
Turbine engines can be ignited readily in ideal atmospheric conditions, but since they often operate in the low temperatures of high altitudes, it is imperative that the system be capable of supplying a high-heat-intensity spark. To supply this spark, a capacitor-type ignition system is utilized.
REFERENCE: ITP POWERPLANT, Chapter 3, Page 44
           AC65-12A, Page 230

4506. ANSWER #4
In a low-tension ignition system, the magneto produces a low voltage which is fed to individual secondary coils located at the spark plugs.
REFERENCE: ITP POWERPLANT, Chapter 3, Page 3
           AC65-12A, Page 186

4507. ANSWER #3
The speed of a magneto with an uncompensated cam is calculated by dividing the number of cylinders by twice the number of poles on the magnet. The speed of the rotating magnet would be 9 divided by 8, or 1 1/8 crankshaft speed. The breaker cam and the rotating magnet both rotate at the same speed, because they are attached to each other, which is 1 1/8 crankshaft speed. Because the distributors only need to fire all plugs in two revolutions of the engine, they rotate at 1/2 crankshaft speed.
REFERENCE: AIRCRAFT POWERPLANTS, Pages 224 and 225

4508. ANSWER #3
The left magneto of a radial engine supplies the spark for the rear plugs in the cylinders. The firing order of a 14 cylinder radial engine is 1, 10, 5, 14, 9, 4, 13, 8, 3, 12, 7, 2, 11, 6. The number 8 ignition lead would fire the 8th cylinder in the order, or cylinder number 8.
REFERENCE: ITP POWERPLANT, Chapter 1, Page 25
           AC65-12A, Pages 19, 182 and 185

4509. ANSWER #2
Spark plugs being gapped too wide will result in engine hard starting. The wide gap settings will raise the "coming in speed" of the magneto, making it hard for the magneto to produce spark at low RPM.
REFERENCE: AC65-12A, Page 216

4510. ANSWER #4
A spark plug is normally loosened from the cylinder bushing by the use of a six-point deep socket. The socket must be seated securely on the spark-plug hexagon to avoid possible damage to the insulator or connector threads.
REFERENCE: ITP POWERPLANT, Chapter 3, Page 34
           AIRCRAFT POWERPLANTS, Page 253

4511. ANSWER #1

Electrical current, when it is finding a path to ground, will take the path of least resistance. If a distributor rotor in a magneto is cracked, the current which is on its way to fire the spark plug may find a better path through the crack to the metal shaft of the magneto.
REFERENCE: AC65-12A, Page 182
           AIRCRAFT POWERPLANTS, Pages 257 and 258

4512. ANSWER #3

The gas turbine engine ignition system is a capacitor discharge type, which produces a high-voltage, high-energy spark.
REFERENCE: ITP POWERPLANT, Chapter 3, Page 34
           AC65-12A, Page 230

4513. ANSWER #2

A high-voltage capacitor discharge ignition system in a turbine engine produces the high voltage at what is called a trigger transformer. This is the second transformer in the system. The first transformer supplies energy to a storage capacitor, and it is the energy coming from the storage capacitor which powers the trigger transformer.
REFERENCE: ITP POWERPLANT, Chapter 3, Page 47
           AC65-12A, Page 232

4514. ANSWER #4

Properly operating points have a fine-grained, frosted, or silvery appearance, while a faulty condenser action produces a coarse-grained and sooty appearance.
REFERENCE: AC65-12A, Page 221

4515. ANSWER #1

Spark plug wires are normally connected to the distributor block with cable-piercing screws.
REFERENCE: AC65-12A, Page 211

4516. ANSWER #4

Cylinder head temperature readings are usually taken at the rear of the cylinder or the rear-most cylinders because this is where the highest temperatures will be encountered.
REFERENCE: AC65-12A, Page 434

4517. ANSWER #2

In most high voltage d.c. capacitor discharge ignition systems, the contacts are opened by a magnetic force and closed by spring action.
REFERENCE: AIRCRAFT GAS TURBINE ENGINE TECHNOLOGY, Page 256

4518. ANSWER #3

Radio noise filters in a turbine engine ignition system operate by blocking the radio frequency noise pulses and shunting them to ground.
REFERENCE: AIRCRAFT GAS TURBINE ENGINE TECHNOLOGY, Page 258

4519. ANSWER #3

Any time a majority of the ignition leads show, under test, an excessive current leakage, the fault may be dirty or improperly treated distributor blocks. If this is the case, the distributor blocks can be cleaned or waxed or, if necessary, replaced.
REFERENCE: AC65-12A, Page 228

4520. ANSWER #1

The electrode gap of the typical turbine engine igniter plug is designed much larger than that of a spark plug, since the operating pressures are much lower and the spark can arc more easily than is the case for a spark plug. Electrode fouling, so common to the spark plug, is minimized by the heat of the high-intensity spark.
REFERENCE: ITP POWERPLANT, Chapter 3, Page 49
           AC65-12A, Page 234

4521. ANSWER #3

In a constrained-gap igniter plug, the center electrode is recessed into the body of the plug. For the spark to get from the electrode to ground, it must jump out away from the plug's tip and then back to ground. Because of this feature, the plug can be out of the flame zone in the combustor, and therefore operate at a cooler temperature.
REFERENCE: AC65-12A, Page 234

4522. ANSWER #2

When inspecting a magneto, all accessible condensers should be cleaned with a lint-free cloth moistened with acetone. Condensers usually have a protective coating on them, a coating which could be damaged by some cleaners but is not damaged by acetone.
REFERENCE: AC65-12A, Page 223

4523. ANSWER #3
Although it is not common in modern systems, some turbine engine ignition systems have used d.c. motors to drive the cam which operates the breaker points. In this type of system, the RPM of the motor is what controls the spark rate at the igniter plug.
REFERENCE: AC65-12A, Page 233

4524. ANSWER #3
The electrode gap of the typical turbine engine igniter plug is designed much larger than that of a spark plug, since the operating pressures are much lower and the spark can arc more easily than is the case for a spark plug. Electrode fouling, so common to the spark plug, is minimized by the heat of the high-intensity spark.
REFERENCE: ITP POWERPLANT, Chapter 3, Page 49
            AC65-12A, Page 234

4525. ANSWER #4
The high-energy current used to fire the turbine engine igniter, if used continuously, would quickly cause electrode erosion. This is not a serious problem, however, because the igniter plugs are only used for short periods of time. Because combustion is continuous in a turbine engine, the ignition system is only needed to get the engine running.
REFERENCE: AC65-12A, Pages 233 and 234

4526. ANSWER #3
Gapping of spark plugs normally consists of decreasing the clearance between the center electrode and the ground electrode. It is not recommended that a gap be widened if it has been inadvertently closed too much, as damage to the nose ceramic or to the electrode can result.
REFERENCE: ITP POWERPLANT, Chapter 3, Page 38

4527. ANSWER #1
With an ignition analyzer, the vertical deflection of the electron beam in the cathode-ray tube is controlled by the voltages present in the primary of the ignition system being tested.
REFERENCE: POWERPLANTS FOR AEROSPACE VEHICLES, Page 270

4528. ANSWER #4
The sparking order of a distributor is always 1 through 9, or 1 through 14, etc., according to the number of cylinders the engine has. The sparking order of the distributor has nothing to do with the firing order of the engine.
REFERENCE: ITP POWERPLANT, Chapter 3, Page 11
            AC65-12A, Page 183

4529. ANSWER #3
The primary capacitor in a magneto must have the correct capacitance. Too low a capacitance will permit arcing and burning of the breaker points plus a weakened output from the secondary.
REFERENCE: ITP POWERPLANT, Chapter 3, Page 10
            AC65-12A, Page 179
            AIRCRAFT POWERPLANTS, Page 227

4530. ANSWER #2
The ignition switch for a magneto ignition system shuts the engine down by grounding out the primary leads to the magnetos. If the primary lead to the right magneto is grounded, it means the right magneto is not producing any secondary current. When the ignition switch is moved to the "right" position, the engine will stop running. In the "both" position and in the "left" position, the engine will be running on only the left magneto.
REFERENCE: ITP POWERPLANT, Chapter 3, Page 24
            AC65-12A, Page 211

4531. ANSWER #2
As mentioned in question #4530, the ignition switch for a magneto ignition sytem shuts the engine down by grounding out the primary leads to the magnetos. If the engine does not shut down when the ignition switch is turned to the "off" position, there is probably an open in the primary leads back to the ignition switch.
REFERENCE: ITP POWERPLANT, Chapter 3, Page 24
            AC65-12A, Page 211

4532. ANSWER #2
In a low-tension ignition system used on an 18 cylinder radial engine, such as the Pratt & Whitney R-2800, there are two distributor assemblies which house four nine-lobe compensated breaker cams (two each). The breakers are mounted adjacent to the cams and are secured to a breaker plate mounted on the distributor housing.
REFERENCE: AIRCRAFT POWERPLANTS, Page 245

4533. ANSWER #1
The high-energy current used to fire the turbine engine igniter, if used continuously, would quickly cause electrode erosion. This is not a serious problem, however, because the igniter plugs are only used for short periods of time. Because combustion is continuous in a turbine engine, the ignition system is only needed to get the engine running.
REFERENCE: AC65-12A, Pages 233 and 234

4534. ANSWER #1
The ignition switch for a magneto ignition system shuts the engine down when turned to the "off" position by grounding out the primary circuits of both magnetos.
REFERENCE: ITP POWERPLANT, Chapter 3, Page 4
AC65-12A, Page 184

4535. ANSWER #4
The firing order of a 14 cylinder radial engine is 1, 10, 5, 14, 9, 4, 13, 8, 3, 12, 7, 2, 11, 6. The number 7 ignition lead would fire the 7th cylinder in the order, or cylinder number 13.
REFERENCE: ITP POWERPLANT, Chapter 1, Page 25
AC65-12A, Pages 19, 182 and 185

4536. ANSWER #4
The left magneto of a radial engine fires the rear plugs of the cylinders.
REFERENCE: AC65-12A, Page 185

4537. ANSWER #2
The operating temperature of a spark plug is controlled through the rate of heat transfer from the plug to the cylinder head.
REFERENCE: ITP POWERPLANT, Chapter 3, Pages 29 and 30
AC65-12A, Page 196

4538. ANSWER #1
In a low-tension ignition sytem, each spark plug is fired by its own secondary coil. If one of the coils fails, one spark plug will be inoperative.
REFERENCE: ITP POWERPLANT, Chapter 3, Page 3
AC65-12A, Page 186

4539. ANSWER #1
When staggered ignition timing is used, the spark plug nearest the exhaust valve fires first. This is done because the fuel/air mixture nearest the exhaust valve is diluted and therefore is slower burning.
REFERENCE: AIRCRAFT POWERPLANTS, Page 221

4540. ANSWER #2
The reach of a spark plug is the length of the threads on the shell, minus any space taken up by the gasket. In essence, it is how far into the cylinder head the plug will reach.
REFERENCE: ITP POWERPLANT, Chapter 3, Pages 27 and 28
AC65-12A, Page 196

4541. ANSWER #2
The right magneto on a radial engine fires the front plugs in the cylinders. The firing order of a 9 cylinder radial engine is 1, 3, 5, 7, 9, 2, 4, 6, 8. Since the question refers to the number 5 position on the distributor block of the right magneto, the plug lead should go to the front plug in the cylinder which fires fifth, which would be #9.
REFERENCE: AC65-12A, Pages 182 and 185

4542. ANSWER #1
The numbers on the distributor block indicate the sparking order of the distributor. These numbers have nothing to do with the firing order of the engine.
REFERENCE: AC65-12A, Page 182

4543. ANSWER #2
After the turbine engine is started, the ignition system can be turned off. Unlike the recip engine, combustion is continuous in a turbine engine and there is no timed ignition.
REFERENCE: ITP POWERPLANT, Chapter 3, Page 44
AC65-12A, Page 230

4544. ANSWER #1
In addition to being able to check ignition leads, a high-tension harness tester can also be used to indicate the condition of the distributor block. If the majority of the ignition leads being tested show excessive leakage, there is a good possibility that the distributor block is at fault.
REFERENCE: AC65-12A, Pages 227 and 228

4545. ANSWER #2
The most likely cause of fractured or broken insulator core tips is the use of improper gapping procedures. If several plugs in the engine are found with this problem, it certainly points toward improper gapping.
REFERENCE: ITP POWERPLANT, Chapter 3, Page 38

4546. ANSWER #4
The breaker-actuating cam in a magneto may be directly driven by the magneto rotor shaft or through a gear train from the rotor shaft. Most large radial engines use a compensated cam, which is designed to operate with a specific engine and has one lobe for each cylinder to be fired. The compensated cam causes the cylinders to fire when the piston position is correct, regardless of the travel of the crankshaft required to obtain that position.
REFERENCE: ITP POWERPLANT, Chapter 3, Page 17
                        AC65-12A, Page 181

4547. ANSWER #1
A spark plug's reach does not relate to its heat characteristics.
REFERENCE: AIRCRAFT POWERPLANTS, Page 252

4548. ANSWER #2
The heat range of a spark plug refers to the ability of the insulator and center electrode to conduct heat away from the tip. A cold spark plug is used in a hot-running, high-compression engine, and a hot spark plug is used in an engine whose cylinder temperatures are reasonably low.
REFERENCE: ITP POWERPLANT, Chapter 3, Page 29

4549. ANSWER #2
When an engine's spark plug is being fired, the current from the secondary is finding a path to ground across the air gap of the spark plug. If the spark plug lead becomes grounded, the current is that much happier because it has a much easier path to ground. The magneto will not be affected by the grounded spark plug lead.
REFERENCE: ITP POWERPLANT, Chapter 3, Pages 2, 3 and 4

4550. ANSWER #1
When the ignition switch for a magneto is in the "both" position, the primary leads to the magnetos are open. The primary leads are grounded when the ignition switch is turned to the "off" position.
REFERENCE: AC65-12A, Page 184

4551. ANSWER #3
The heat range of a spark plug is a measure of its ability to transfer heat to the cylinder head. The length of the nose core is the principal factor in establishing the plug's heat range. "Hot" plugs have a long insulator nose that creates a long heat transfer path, whereas "cold' plugs have a relatively short insulator to provide a rapid transfer of heat to the cylinder head.
REFERENCE: ITP POWERPLANT, Chapter 3, Page 29
                        AC65-12A, Page 196

4552. ANSWER #4
In a battery ignition system, current flows through the primary of the coil any time the ignition switch is turned on and the breaker points are closed.
REFERENCE: ITP POWERPLANT, Chapter 3, Page 2

4553. ANSWER #2
In a magneto ignition sytem, the purpose of the ignition switch is to ground out the primary of the magneto when the switch is "off", and to open the primary lead to the magneto when the switch is "on".
REFERENCE: ITP POWERPLANT, Chapter 3, Page 4
                        AC65-12A, Page 184

4554. ANSWER #4
When performing a magneto ground check, the engine is operated at a specified RPM and the ignition switch is moved from the "both" postion to "left" and the "both" position to "right". When this check is performed there should be a small drop in RPM at the "left" and the "right" positions. This is normal because in each case one of the spark plugs in the cylinder is not operating.
REFERENCE: ITP POWERPLANT, Chapter 3, Page 24
                        AC65-12A, Pages 437, 439 and 440

4555. ANSWER #4
A defective spark plug, meaning a plug which is not firing, will cause the engine to miss at all engine speeds. The cylinder which contains the defective plug will produce less power than the other cylinders, and it will affect the engine throughout the entire RPM range.
REFERENCE: AC65-12A, Page 456

4556. ANSWER #4
A spark plug is fouled when it becomes contaminated with foreign matter to the point that the spark flows through the foreign matter to ground rather than jumping the air gap at the electrode.
REFERENCE: ITP POWERPLANT, Chapter 3, Page 35
              AC65-12A, Page 216
              AIRCRAFT TECHNICAL DICTIONARY, Page 104

4557. ANSWER #2
A spark plug whose insulator tip is cracked should be replaced. This condition can affect the firing of the plug and also its ability to transfer heat.
REFERENCE: ITP POWERPLANT, Chapter 3, Page 36

4558. ANSWER #3
Fundamentally, an engine which runs hot requires a relatively cold spark plug, whereas an engine which runs cool requires a relatively hot spark plug. If a hot spark plug is installed in an engine which runs hot, the tip of the plug will be overheated and cause pre-ignition.
REFERENCE: AIRCRAFT POWERPLANTS, Page 252

4559. ANSWER #2
Carbon fouling of spark plugs from fuel is associated with mixtures that are too rich to burn. A rich fuel/air mixture is detected by soot or black smoke coming from the exhaust, by an increase in RPM when the idling fuel/air mixture is leaned to "best power", and by soot collecting on the tip of the spark plug.
REFERENCE: ITP POWERPLANT, Chapter 3, Page 36
              AC65-12A, Page 213

4560. ANSWER #2
The heat range of a spark plug is a measure of its ability to transfer heat to the cylinder head.
REFERENCE: ITP POWERPLANT, Chapter 3, Page 29
              AC65-12A, Page 196

4561. ANSWER #1
When performing a magneto check on an engine, fast RPM drop is usually the result of faulty spark plugs or faulty ignition harness. This is true because faulty plugs or leads take effect at once.
REFERENCE: ITP POWERPLANT, Chapter 3, Page 24
              AC65-12A, Page 440

4562. ANSWER #1
When installing new breaker points in a magneto, the internal timing of the magneto must be checked to ensure that the point opening coincides with the E-gap position of the rotor. The timing of the magneto to the engine should also be checked to ensure that the firing of the spark plugs will occur at the proper time in relation to piston position.
REFERENCE: ITP POWERPLANT, Chapter 3, Pages 17 and 18
              AC65-12A, Page 203
              AC43.13-1A, Page 280

4563. ANSWER #1
A high compression engine operates with high combustion chamber temperatures. To operate in this environment, a "cold" spark plug should be installed.
REFERENCE: ITP POWERPLANT, Chapter 3, Page 29

4564. ANSWER #2
Lead fouling may occur at any power setting, but the power setting most conducive to lead fouling is cruising with lean mixtures. At this power, the cylinder head temperature is relatively low and there is an excess of oxygen over that needed to consume all the fuel in the fuel/air mixture. This oxygen ends up combining with the lead and builds up in layers on the cool cylinder walls and the spark plugs.
REFERENCE: ITP POWERPLANT, Chapter 3, Page 35
              AC65-12A, Page 214

4565. ANSWER #2
The ignition event in a four-stroke engine occurs before the piston reaches top dead center on the compression stroke.
REFERENCE: ITP POWERPLANT, Chapter 1, Page 8
              AC65-12A, Page 29

4566. ANSWER #2
When timing a magneto to an engine, the first thing that must be known is that the magneto is internally timed. This being the case, the engine must then be prepared to accept the installation of the magneto. With the magneto locked in position, the engine is rotated until the piston in the number one cylinder is the prescribed number of degrees before top dead center on the compression stroke. When this position is obtained, the magneto can be installed.
REFERENCE: ITP POWERPLANT, Chapter 3, Page 23
AC65-12A, Page 203

4567. ANSWER #4
Whether the ignition system is a battery or a magneto type, the spark occurs at the spark plug when the primary circuit is broken. The opening of the breaker points is what breaks the primary circuit.
REFERENCE: ITP POWERPLANT, Chapter 3, Pages 9 and 10
AC65-12A, Page 179

4568. ANSWER #4
Because of its high-energy spark, the capacitor discharge ignition system is the type used most often on turbine engines.
REFERENCE: ITP POWERPLANT, Chapter 3, Page 44
AC65-12A, Page 230

4569. ANSWER #3
When performing a magneto check on an aircraft engine, a slow drop in RPM is usually caused by incorrect ignition timing or faulty valve adjustment. With late ignition timing, the charge is fired too late in relation to piston travel for the combustion pressures to build up to the maximum. Incorrect valve clearances, through their effect on valve overlap, can cause the mixture to be too rich or too lean.
REFERENCE: ITP POWERPLANT, Chapter 3, Page 24
AC65-12A, Page 440

4570. ANSWER #4
In a magneto ignition system, the purpose of the ignition switch is to ground out the primary of the magneto when the switch is "off", and to open the primary lead to the magneto when the switch is "on". If the ground wire to the ignition switch is disconnected, there is no way for the switch to ground out the primary of the magneto, and therefore no way to shut off the engine.
REFERENCE: ITP POWERPLANT, Chapter 3, Page 4
AC65-12A, Page 184

4571. ANSWER #4
The principal advantages of the dual magneto-ignition system are the following:
1. If one magneto or any part of one magneto system fails to operate, the other magneto will furnish ignition until the disabled system functions again.
2. Two sparks, igniting the fuel/air mixture in each cylinder simultaneously at two different places, give a more complete and quick combustion than a single spark; hence, the power of the engine is increased.
REFERENCE: ITP POWERPLANT, Chapter 3, Page 1
AIRCRAFT POWERPLANTS, Page 220

4572. ANSWER #4
The ignition harness and shielding serve as a conductor for stray magnetic fields that surround the wires as they momentarily carry high-voltage current. By conducting these magnetic lines of force to ground, the ignition harness cuts down electrical interference with the aircraft radio and other electrically sensitive equipment.
REFERENCE: ITP POWERPLANT, Chapter 3, Page 42
AC65-12A, Page 184

4573. ANSWER #1
The high-tension magneto system can be divided into three distinct circuits: the magnetic, the primary electrical, and the secondary electrical.
REFERENCE: ITP POWERPLANT, Chapter 3, Page 6
AC65-12A, Page 177

4574. ANSWER #1
The distributor in a magneto ignition system is made up of two parts. The revolving part is called a distributor rotor and the stationary part is called a distributor block.
REFERENCE: AC65-12A, Page 182

4575. ANSWER #4
Flashover in a distributor can lead to carbon tracking, which appears as a fine pencil-like line on the unit across which flashover occurs. The carbon trail results from the electrical spark burning dirt particles which contain hydrocarbon materials.
REFERENCE: AC65-12A, Page 183

4576. ANSWER #2
The distributor of a magneto ignition system turns at 1/2 crankshaft speed. Because it takes two revolutions of the engine to fire all the cylinders, the distributor only needs to rotate 1/2 as fast.
REFERENCE: AC65-12A, Page 182

4577. ANSWER #3
The fuel/air mixture in turbine engines can be ignited readily in ideal atmospheric conditions, but since they often operate in the low temperatures of high altitudes, it is imperative that the system be capable of supplying a high heat intensity spark.
REFERENCE: AC65-12A, Page 230

4578. ANSWER #4
A typical turbine engine ignition system includes two exciter units, two transformers, two intermediate ignition leads, and two high-tension leads.
REFERENCE: AC65-12A, Page 230

4579. ANSWER #2
An ignition switch check is usually made at 700 RPM. On those aircraft engine installations that will not idle at this low RPM, the engine speed should be set to the minimum possible.
REFERENCE: AC65-12A, Page 211

4580. ANSWER #2
The primary breaker points in a magneto close at approximately the full register position. In this position, the poles of the magnet are perfectly aligned with the pole shoes.
REFERENCE: ITP POWERPLANT, Chapter 3, Page 8
            AC65-12A, Page 179

4581. ANSWER #3
A dual magneto is one that incorporates two magnetos in one housing. One rotating magnet and a cam are common to two sets of breaker points and coils.
REFERENCE: ITP POWERPLANT, Chapter 3, Page 5
            AC65-12A, Page 185

4582. ANSWER #3
If the breaker points in a magneto are spread wider than recommended, it is possible for the mainspring to take a permanent set. If this happens, the spring can lose some of its closing tension. The end result will be points that tend to "bounce" or "float".
REFERENCE: AC65-12A, Page 220

4583. ANSWER #3
The secondary coil of a magneto is made up of a winding containing approximately 13,000 turns of fine, insulated wire, one end of which is electrically grounded to the primary coil or to the coil core and the other end connected to the distributor rotor.
REFERENCE: AC65-12A, Page 181

4584. ANSWER #2
A typical water injection system includes a tank fitted with an electrically operated pump which delivers water under pressure to a nonhesitating-type water regulator. A switch actuated by engine oil pressure ensures that the pump cannot be operated unless the engine is running.
REFERENCE: AC65-12, Page 148 (Appears in the 65-12, not the 65-12A)

4585. ANSWER #4
A carburetor equipped with a derichment valve and a derichment jet is part of an anti-detonant injection system.
REFERENCE: AC65-12, Page 148 (Appears in the 65-12, not the 65-12A)

4586. ANSWER #4
In a turbine engine water injection system, water from the aircraft tank system is routed to two shutoff valves which govern flow to the two water injection controls. The shutoff valves are armed by the actuation of a cockpit switch. The selected valve opens or closes upon receipt of an electrical signal from the fuel control water injection switch.
REFERENCE: ITP POWERPLANT, Chapter 6, Page 72
            AC65-12A, Pages 174 and 175

4587. ANSWER #2

In recip engines, the anti-detonant fluid used in water injection systems is alcohol and water. In turbine engines, pure demineralized water is used.
REFERENCE: AC65-12A, Page 147

4588. ANSWER #3

The derichment valve in the carburetor will only operate if water is flowing. Because water cannot flow if the engine is not running (no oil pressure), the derichment valve will not operate.
REFERENCE: AC65-12, Page 148 (Appears in the 65-12, not the 65-12A)

4589. ANSWER #3

At high power settings, the carburetor delivers more fuel to the engine than it needs to help keep it cool. A leaner mixture would produce more power, but then overheating would be a problem. Water injection allows the engine to operate a leaner mixture because of the cooling effect of the water. Operating on this best power mixture, the engine develops more power even though the manifold pressure and RPM settings remain unchanged.
REFERENCE: AC65-12A, Page 147

4590. ANSWER #2

At high power settings, the carburetor delivers more fuel to the engine than it needs to help keep it cool. A leaner mixture would produce more power, but then overheating would be a problem. Water injection allows the engine to operate with a leaner mixture because of the cooling effect of the water. Operating on this best power mixture, the engine develops more power even though the manifold pressure and RPM settings remain unchanged.
REFERENCE: AC65-12A, Page 147

4591. ANSWER #3

When the ADI fluid starts flowing in a recip engine, it supplies a pressure signal to the diaphragm-operated derichment valve in the carburetor telling it to lean the fuel/air mixture to approximately the "best power" setting.
REFERENCE: AC65-12, Page 150 (Appears in the 65-12, not the 65-12A)

4592. ANSWER #3

The automatic mixture control unit on many modern carburetors is designed to alter the fuel flow to compensate for changes in air density due to temperature and altitude changes. The density of the air is a measure of its mass.
REFERENCE: ITP POWERPLANT, Chapter 6, Page 41
             AC65-12A, Pages 127 and 128

4593. ANSWER #4

For an engine to develop maximum power at full throttle, the fuel mixture must be richer than for cruise. The additional fuel is used for cooling the engine to prevent detonation. The economizer valve is the one which adds the extra fuel at high power settings.
REFERENCE: ITP POWERPLANT, Chapter 6, Page 20
             AC65-12A, Page 121

4594. ANSWER #1

The discharge nozzle in a float-type carburetor is located in the throat of the venturi at the point where the lowest drop in pressure occurs. Thus, there are two different pressures acting on the fuel in the carburetor — a low pressure at the discharge nozzle and a higher pressure acting in the float chamber.
REFERENCE: ITP POWERPLANT, Chapter 6, Page 15
             AC65-12A, Page 115

4595. ANSWER #2

The main air bleed in a float-type carburetor allows air to be drawn in along with the fuel at the venturi of the carb. This air bleed helps decrease the fuel density and destroy surface tension. This results in better vaporization and control of fuel discharge, especially at lower engine speeds. If the air bleed becomes clogged, the engine will draw too much fuel at rated power and run rich.
REFERENCE: ITP POWERPLANT, Chapter 6, Page 16
             AC65-12A, Page 119

4596. ANSWER #2

Most carburetors have the float level adjusted by adding or removing shims between the needle seat and the throttle body.
REFERENCE: ITP POWERPLANT, Chapter 6, Page 25

4597. ANSWER #2

The automatic mixture control on a pressure carburetor contains a metallic bellows with a pressure of 28 inches Hg sealed inside it. As the density of the air changes, the expansion and contraction of the bellows moves a tapered needle in the atmospheric line. As the aircraft climbs and the atmospheric pressure decreases, the bellows expands, inserting the tapered needle farther and farther into the atmospheric passage and restricting the flow of air to chamber "A" of the regulator unit. This causes the regulator to decrease the fuel flow to the engine.
REFERENCE: AC65-12A, Page 128

4598. ANSWER #2

The main air bleed in a float-type carburetor allows air to be drawn in along with the fuel at the venturi of the carb. This air bleed helps decrease the fuel density and destroy surface tension. This results in better vaporization and control of fuel discharge, especially at lower engine speeds. If the air bleed becomes clogged, the engine will draw too much fuel at rated power and run rich.
REFERENCE: ITP POWERPLANT, Chapter 6, Page 16
             AC65-12A, Page 119

4599. ANSWER #2

The main bleed in a float-type carburetor allows air to be drawn in along with the fuel at the venturi of the carb. This air bleed helps decrease the fuel density and destroy surface tension. This results in better vaporization and control of fuel discharge, especially at lower engine speeds. If the air bleed becomes clogged, the engine will draw too much fuel at rated power and run rich. The engine will run rich because the suction through the venturi will act with more force on the fuel discharge nozzle, now that it can't act on the air bleed.
REFERENCE: ITP POWERPLANT, Chapter 6, Page 16
             AC65-12A, Page 119

4600. ANSWER #2

During the operation of a carburetor, the float assumes a position slightly below its highest level to allow a valve opening sufficient for replacement of the fuel being consumed by the engine. If the fuel level in the float chamber is too high, the mixture will be rich. Because a punctured float will fill with fuel, it may not be able to float at all or will float at a lower level. In either case, the level of fuel in the float chamber will be higher than normal and the mixture will be rich, possibly to the point of flooding out the engine.
REFERENCE: AC65-12A, Page 117
             AIRCRAFT POWERPLANTS, Page 79

4601. ANSWER #3

The back-suction type mixture control system is the most widely used on float-type carburetors. In this system a certain amount of venturi low pressure acts upon the fuel in the float chamber so that it opposes the low pressure existing at the main discharge nozzle. By varying the pressure acting on the fuel in the float chamber, the system varies the mixture being supplied to the engine.
REFERENCE: ITP POWERPLANT, Chapter 6, Page 17
             AC65-12A, Page 121

4602. ANSWER #3

The low pressure area created by the venturi of a carburetor is dependent upon air velocity rather than air density. The action of the venturi draws the same volume of fuel through the discharge nozzle at a high altitude as it does at a low altitude. Therefore, the fuel mixture becomes richer as altitude increases. This can be overcome by either a manual or an automatic mixture control.
REFERENCE: AC65-12A, Page 120

4603. ANSWER #2

An economizer system is designed to enrich the fuel mixture when the throttle is advanced to full power. The economizer allows the mixture to be adjusted to a lean setting for most engine operations, because the mixture will be automatically enriched at full power.
REFERENCE: ITP POWERPLANT, Chapter 6, Page 20
             AC65-12A, Pages 121 and 122

4604. ANSWER #1

A likely cause of an engine flooding is a float bowl with too high a fuel level. This can be caused by the float level being improperly set or by a leak at the needle valve and seat.
REFERENCE: AIRCRAFT POWERPLANTS, Page 95

4605. ANSWER #1

The main air bleed in a float-type carburetor allows air to be drawn in along with the fuel at the venturi of the carb. This air bleed helps decrease the fuel density and destroy surface tension. This results in better vaporization and control of fuel discharge, especially at lower engine speeds. If the air bleed becomes clogged, the engine will draw too much fuel at rated power and run rich. The engine will run rich because the suction through the venturi will act with more force on the fuel discharge nozzle, now that it can't act on the air bleed.
REFERENCE: ITP POWERPLANT, Chapter 6, Page 16
             AC65-12A, Page 119

4606. ANSWER #2
The back-suction type mixture control system is the most widely used in float-type carburetors. In this system a certain amount of venturi low pressure acts upon the fuel in the float chamber so that it opposes the low pressure existing at the main discharge nozzle. By varying the pressure acting on the fuel in the float chamber, the system varies the mixture being supplied to the engine. When the mixture control is placed in "idle cutoff", the float bowl is connected to a passage which leads to piston suction.
REFERENCE: ITP POWERPLANT, Chapter 6, Page 17
             AC65-12A, Page 121

4607. ANSWER #1
The function of the automatic mixture control is to regulate the richness of the fuel/air charge entering the engine.
REFERENCE: ITP POWERPLANT, Chapter 6, Pages 30 and 31
             AC65-12A, Page 128

4608. ANSWER #3
With the throttle valve closed at idling speeds, air velocity through the venturi is so low that it cannot draw enough fuel from the main discharge nozzle. However, low pressure exists on the engine side of the throttle valve. In order to allow the engine to idle, a fuel passageway is incorporated to discharge fuel from an opening in the low pressure area near the edge of the throttle valve. This opening is called the idling jet.
REFERENCE: ITP POWERPLANT, Chapter 6, Page 18
             AC65-12A, Page 120

4609. ANSWER #2
In the pressure-type carburetor, the fuel follows the same path at idling as it does when the main metering system is in operation. Because of the low velocity of the air through the venturi, however, the differential pressure on the air diaphragm is not sufficient to regulate the fuel flow. Instead, the idle spring in the unmetered fuel chamber holds the poppet valve off its seat to admit fuel from the carburetor inlet during operation at idling speeds.
REFERENCE: AC65-12A, Page 130

4610. ANSWER #3
For an engine to develop maximum power at full throttle, the fuel mixture must be richer than for cruise. The additional fuel is used for cooling the engine to prevent detonation. The economizer valve is the one which supplies and regulates the extra fuel at high power settings.
REFERENCE: ITP POWERPLANT, Chapter 6, Page 20
             AC65-12A, Page 121

4611. ANSWER #4
The automatic mixture control on a pressure carburetor contains a metallic bellows with a pressure of 28 inches Hg sealed inside it. As the density of the air changes, the expansion and contraction of the bellows moves a tapered needle in the atmospheric line. As the aircraft climbs and the atmospheric pressure decreases, the bellows expands, inserting the tapered needle farther and farther into the atmospheric passage and restricting the flow of air to chamber "A" of the regulator unit. This causes the regulator to decrease the fuel flow to the engine. If the bellows were to rupture, no decrease in fuel flow would occur and the engine would run rich.
REFERENCE: AC65-12A, Page 128

4612. ANSWER #3
In a float-type carburetor, a float chamber is provided between the fuel supply and the metering system. The float chamber provides a nearly constant level of fuel to the main discharge nozzle. This level is usually about 1/8 inch below the holes in the main discharge nozzle.
REFERENCE: ITP POWERPLANT, Chapter 6, Page 15
             AC65-12A, Page 117

4613. ANSWER #3
In a pressure carburetor system, the drop in pressure at the throat of the boost venturi is proportional to the airflow. This causes a lowering in the pressure in chamber "B". Chamber "A" has impact pressure in it, so the difference in the pressures in the two chambers is a measure of airflow. As the difference in the two chamber pressures changes, the diaphragm separating them moves in the direction of the lesser pressure, causing a poppet valve to move. The moving of the poppet valve is what causes fuel flow to change. If the diaphragm separating chambers "A" and "B" were to rupture, there would be no movement of the poppet valve. Because the poppet valve moves in the direction of increased fuel flow when the throttle is advanced (because of a decrease in chamber "B" pressure), the engine would run lean at all power settings.
REFERENCE: AC65-12A, Page 129

4614. ANSWER #2
In a pressure carburetor system, the drop in pressure at the throat of the boost venturi is proportional to the airflow. This causes a lowering in the pressure in chamber "B". Chamber "A" has impact pressure in it, so the difference in the pressures in the two chambers is a measure of airflow. As the difference in the two chamber pressures changes, the diaphragm separating them moves in the direction of the lesser pressure, causing a poppet valve to move. The moving of the poppet valve is what causes fuel flow to change. If the diaphragm separating chambers "A" and "B" were to rupture, there would be no movement of the poppet valve. Because the poppet valve moves in the direction of increased fuel flow when the throttle is advanced (because of a decrease in chamber "B" pressure), the engine would run lean at all power settings.
REFERENCE: AC65-12A, Page 129

4615. ANSWER #1
Chamber "C" in a pressure carb is filled with fuel at the same pressure as that in the discharge line. The function of chamber "C" is to allow fuel to be discharged under pressure and to compensate for variations of pressures in the discharge line. The discharge nozzle acts as a relief valve to hold this pressure fairly constant regardless of the volume of fuel being discharged.
REFERENCE: AC65-12A, Page 129

4616. ANSWER #2
The pressure at the boost venturi varies according to the airflow through the carburetor. At high airflow conditions, the pressure at the boost venturi and therefore chamber "B", is at its lowest. For this reason, retarding the throttle, which will cause the boost venturi pressure to increase, will decrease the pressure differential between chambers "A" and "B".
REFERENCE: AC65-12A, Page 129

4617. ANSWER #4
When checking the fuel level in the float chamber of a float-type charburetor, it should never be checked at the edge of the chamber. The reason for this is the fuel tends to cling to the walls of the chamber, which will cause it to be a little higher at this point than it is in the center of the chamber.
REFERENCE: ITP POWERPLANT, Chapter 6, Page 26

4618. ANSWER #4
In the pressure-type carburetor, the fuel follows the same path at idling as it does when the main metering system is in operation. Because of the low velocity of the air through the venturi, however, the differential pressure on the air diaphragm is not sufficient to regulate the fuel flow. Instead, the idle spring in the unmetered fuel chamber holds the poppet valve off its seat to admit fuel from the carburetor inlet during operation at idling speeds
REFERENCE: AC65-12A, Page 130

4619. ANSWER #4
The carburetor must measure the airflow through the induction system and use this measurement to regulate the amount of fuel discharged into the airstream. The air measuring unit is the venturi, which makes use of a basic law of physics known as Bernoulli's theorem.
REFERENCE: ITP POWERPLANT, Chapter 6, Page 14
                       AC65-12A, Page 113

4620. ANSWER #1
The automatic mixture control on a pressure carburetor contains a metallic bellows with a pressure of 28 inches Hg sealed inside it. As the density of the air changes, the expansion and contraction of the bellows moves a tapered needle in the atmospheric line. As the aircraft climbs and the atmospheric pressure decreases, the bellows expands, inserting the tapered needle farther and farther into the atmospheric passage and restricting the flow of air to chamber "A" of the regulator unit. This causes the regulator to decrease the fuel flow to the engine.
REFERENCE: AC65-12A, Page 128

4621. ANSWER #1
If an aircraft engine is equipped with carburetor heat, it should be started with the heat control placed in the "cold" position. This is done to prevent damage and possible fire in case the engine backfires.
REFERENCE: AC65-9A, Page 490

4622. ANSWER #2
Idling on a float-type carburetor is accomplished by allowing fuel to flow from an idle jet or idle discharge nozzle. The main discharge nozzle cannot be used at idle speeds because there is not sufficient airflow.
REFERENCE: ITP POWERPLANT, Chapter 6, Page 18
                       AC65-12A, Page 120

4623. ANSWER #4
Bernoulli's theorem and the general gas law can be used to explain what happens to air as it flows through the carburetor's venturi. Bernoulli's theorem says that when air flows through a converging duct (carburetor venturi), the velocity of the air goes up and the pressure of the air goes down. The general gas law identifies that pressure and temperature are directly proportional. So a decrease in pressure means a decrease in temperature.
REFERENCE: ITP POWERPLANT, Chapter 6, Page 14
             AC65-9A, Page 236
             AC65-12A, Page 124

4624. ANSWER #2
In a float-type carburetor, the throttle valve is located downstream of the main discharge nozzle and the venturi. It is the throttle valve which controls the mass airflow through the venturi, so it needs to be downstream of it.
REFERENCE: ITP POWERPLANT, Chapter 6, Page 14 (Figure 1-6)
             AC65-12A, Page 118 (Figure 3-8)

4625. ANSWER #4
The float level in a float-type carburetor is determined by measuring the distance from the top of the fuel to the parting surface of the carburetor body.
REFERENCE: AIRCRAFT POWERPLANTS, Page 94

4626. ANSWER #3
The low pressure area created by the venturi of a carburetor is dependent upon air velocity rather than air density. The action of the venturi draws the same volume of fuel through the discharge nozzle at a high altitude as it does at a low altitude. Therefore, the fuel mixture becomes richer as altitude increases. This can be overcome either by a manual or an automatic mixture control.
REFERENCE: AC65-12A, Page 120

4627. ANSWER #3
The venturi in a carburetor performs three functions: (1) Proportions the fuel/air mixture, (2) decreases the pressure at the discharge nozzle, and (3) limits the airflow at full throttle.
REFERENCE: AC65-12A, Page 118

4628. ANSWER #1
The back-suction type mixture control system is the most widely used in float-type carburetors. In this system a certain amount of venturi low pressure acts upon the fuel in the float chamber so that it opposes the low pressure existing at the main discharge nozzle. By varying the pressure acting on the fuel in the float chamber, the system varies the mixture being supplied to the engine. When the mixture control is placed in "idle cutoff", the float bowl is connected to a passage which leads to piston suction.
REFERENCE: ITP POWERPLANT, Chapter 6, Page 17
             AC65-12A, Page 121

4629. ANSWER #1
The main air bleed on a float-type carburetor allows air to be mixed with the fuel being drawn out of the main discharge nozzle. Air bled into the main metering fuel system decreases the fuel density and destroys surface tension. This results in better vaporization and control of fuel discharge, especially at lower engine speeds.
REFERENCE: AC65-12A, Page 120

4630. ANSWER #1
The float level in a float-type carburetor is determined by measuring the distance from the top of the fuel to the parting surface of the carburetor body.
REFERENCE: AIRCRAFT POWERPLANTS, Page 94

4631. ANSWER #3
The throttle valve on a float-type carburetor is located between the venturi and the engine.
REFERENCE: ITP POWERPLANT, Chapter 6, Page 14 (Figure 1-6)
             AC65-12A, Page 118 (Figure 3-8)

4632. ANSWER #3
The purpose of the automatic mixture control is to compensate for changes in air density due to temperature and altitude changes. These changes are compensated for by varying the amount of fuel flowing to the engine. Also see the explanation for question #4620.
REFERENCE: AC65-12A, Page 128

4633. ANSWER #1
An increased amount of fuel (richer mixture) is used in the idle range because at idling speeds, the engine may not have enough air flowing around its cylinders to provide proper cooling.
REFERENCE: AIRCRAFT POWERPLANTS, Page 80

4634. ANSWER #1
When the throttle valve is opened quickly, a large volume of air rushes through the air passage of the carburetor. The amount of fuel which is mixed with this air, however, is less than normal. This causes a momentary leaning of the mixture. To overcome this tendency, the carburetor is equipped with a small fuel pump called an accelerating pump.
REFERENCE: ITP POWERPLANT, Chapter 6, Pages 19 and 20
            AC65-12A, Page 121

4635. ANSWER #2
A mixture control on a carburetor is the device which controls the ratio of the fuel/air charge being burned in the cylinders. Depending on the type of carburetor, the mixture control can be a manual or an automatic device.
REFERENCE: ITP POWERPLANT, Chapter 6, Page 17
            AC65-12A, Pages 120 and 121

4636. ANSWER #4
The throttle valve controls the mass airflow through the carburetor venturi by acting as a variable restriction. It is located downstream of the venturi and the main discharge nozzle.
REFERENCE: ITP POWERPLANT, Chapter 6, Page 16
            AC65-12A, Page 118

4637. ANSWER #2
Some continuous-flow fuel injection systems used on aircraft engines have a fuel discharge nozzle located in each cylinder head. The nozzle outlet is directed into the intake port.
REFERENCE: ITP POWERPLANT, Chapter 6, Page 39
            AC65-12A, Page 140

4638. ANSWER #3
According to Bernoulli's theorem, when air flows at a continuous rate, its pressure is indirectly proportional to its velocity. When the air flows through the venturi of the carburetor, its velocity increases and its pressure decreases. The amount of pressure decrease is dependent on the amount of velocity increase.
REFERENCE: ITP POWERPLANT, Chapter 6, Page 14
            AC65-12A, Page 119

4639. ANSWER #3
The accelerator pump in a pressure injection carburetor operates with fuel on one side of a diaphragm and spring pressure and throttle valve pressure (partial vacuum) on the other side. When the engine is operating with a steady throttle setting, the fuel pressure and spring/air pressure are in balance. When the engine throttle is advanced, the suction downstream of the throttle valve decreases (pressure increases) which allows the spring force in the accelerator pump to force fuel into the induction system. Single and double diaphragm accelerator pumps operate the same way, whether the engine is supercharged or not.
REFERENCE: ITP POWERPLANT, Chapter 6, Pages 32 and 33
            AC65-12A, Page 130

4640. ANSWER #2
The accelerator pump in a pressure injection carburetor operates with fuel on one side of a diaphragm and spring pressure and throttle valve pressure (partial vacuum) on the other side. When the engine is operating with a steady throttle setting, the fuel pressure and spring/air pressure are in balance. When the engine throttle is advanced, the suction downstream of the throttle valve decreases (pressure increases) which allows the spring force in the accelerator pump to force fuel into the induction system.
REFERENCE: ITP POWERPLANT, Chapter 6, Pages 32 and 33
            AC65-12A, Page 130

4641. ANSWER #4
Fuel metering in a float-type carburetor is accomplished with one of two jets: the idle jet or the main metering jet. The idle jet functions only at idle, while the main metering jet takes care of all throttle settings above idle.
REFERENCE: ITP POWERPLANT, Chapter 6, Page 18
            AC65-12A, Page 120

4642. ANSWER #4
In a continuous cylinder fuel injection system, there is no timed injection between the injector pump and the cylinders. Fuel is always available at the intake port in each cylinder.
REFERENCE: ITP POWERPLANT, Chapter 6, Page 37
            AC65-12A, Page 140

4643. ANSWER #2
When the throttle valve is opened quickly, a large volume of air rushes through the air passage of the carburetor. The amount of fuel which is mixed with this air, however, is less than normal. This causes a momentary leaning of the mixture. To overcome this tendency, the carburetor is equipped with a small fuel pump called an accelerating pump.
REFERENCE: ITP POWERPLANT, Chapter 6, Paged 19 and 20
             AC65-12A, Page 121

4644. ANSWER #1
The carburetor is mounted on the engine so that air to the cylinders passes through the barrel, the part of the carburetor which contains the venturi. The size and shape of the venturi depends on the requirements of the engine for which the carburetor is designed. A carburetor for a high-powered engine will have a venturi which is larger than the venturi for a lower powered engine.
REFERENCE: AC65-12A, Page 113

4645. ANSWER #1
In a continuous cylinder fuel injection system, there is no timed injection between the injector pump and the cylinders. Fuel is always available at the intake port in each cylinder.
REFERENCE: ITP POWERPLANT, Chapter 6, Page 37
             AC65-12A, Page 140

4646. ANSWER #1
In a pressure injection carburetor, the accelerating pump and the enrichment valve operate independently of each other. The accelerating pump operates on a momentary basis any time the throttle is opened quickly. The enrichment valve operates continuously when the engine is at high power.
REFERENCE: ITP POWERPLANT, Chapter 6, Pages 32 and 33
             AC65-12A, Pages 130 and 133

4647. ANSWER #4
According to Bernoulli's theorem, when air flows at a continuous rate, its pressure is indirectly proportional to its velocity. When the air flows through the venturi of the carburetor, its velocity increases and its pressure decreases. The amount of pressure decrease is dependent on the amount of velocity increase.
REFERENCE: ITP POWERPLANT, Chapter 6, Page 14
             AC65-12A, Page 113

4648. ANSWER #4
Backfiring is a condition where flame is lingering in the cylinder when the intake valve opens. The flame ignites the fuel/air charge coming in through the intake valve, which causes an explosion in the induction system. In a direct cylinder fuel injection system, the intake valve only allows air to enter the cylinder, while the fuel is injected through a separate nozzle. This type of system is not susceptible to backfiring.
REFERENCE: AC65-12A, Pages 136 and 445

4649. ANSWER #4
According to Bernoulli's theorem, when air flows at a continuous rate, its pressure is indirectly proportional to its velocity. When the air flows through the venturi of the carburetor, its velocity increases and its pressure decreases. The amount of pressure decrease is dependent on the amount of velocity increase.
REFERENCE: ITP POWERPLANT, Chapter 6, Page 14
             AC65-12A, Page 113

4650. ANSWER #2
The throttle valve limits the airflow through the carburetor at all throttle settings except full throttle. At full throttle, the venturi limits the airflow.
REFERENCE: ITP POWERPLANT, Chapter 6, Page 16
             AC65-12A, Page 118

4651. ANSWER #1
The low pressure area created by the venturi is dependent upon air velocity rather than air density. The action of the venturi draws the same volume of fuel through the discharge nozzle at a high altitude as it does at a low altitude. Therefore, the fuel mixture becomes richer as altitude increases.
REFERENCE: AC65-12A, Page 120

4652. ANSWER #3
If carburetor heat is applied during engine operation, the temperature of the air entering the carburetor will be increased and its density will be decreased. A reduced mass of air mixing with the same quantity of fuel will cause a rich mixture. The end result will be a decreased air to fuel ratio. Answer #3 says a decreased fuel to air ratio, which is not a technically correct answer. However, on page 112 of the AC65-12A, it is stated as a fuel to air ratio of 12 to 1. In case you are wondering about answer #4, the volume of air through the carburetor will not change, only the mass of air will change.
REFERENCE: AC65-12A, Pages 72 and 112

4653. ANSWER #1
Carburetor adjustments, such as idle speed and mixture, should not be made unless the engine is warmed up and operating in the normal temperature range. Fuel vaporization is different in the engine when it is cold from when it is at operating temperature, so an adjustment made on a cold engine will not be accurate.
REFERENCE: ITP POWERPLANT, Chapter 6, Page 26
              AC65-12A, Page 144

4654. ANSWER #1
Because of the possiblity of liquid lock, the priming system on a radial engine only primes the cylinders from the horizontal up. On a nine cylinder radial engine, this would be cylinders one, two, three, eight, and nine.
REFERENCE: AIRCRAFT PROPULSION POWERPLANTS, Page 158

4655. ANSWER #4
The idling circuit in a float-type carburetor includes a separate idle jet and idle air bleed. The idle air bleed helps the fuel to vaporize before it is drawn into the cylinder.
REFERENCE: ITP POWERPLANT, Chapter 6, Page 18
              AC65-12A, Page 120

4656. ANSWER #4
According to Bernoulli's theorem, when air flows at a continuous rate, its pressure is indirectly proportional to its velocity. When the air flows through the venturi of the carburetor, its velocity increases and its pressure decreases. If the volume of air flowing through the carburetor decreases, the pressure will increase.
REFERENCE: ITP POWERPLANT, Chapter 6, Page 14
              AC65-12A, Page 113

4657. ANSWER #4
The fuel/air ratio of an engine through its operating range runs from very rich at idle, to lean at cruise, to rich at full throttle.
REFERENCE: AC65-12A, Page 110

4658. ANSWER #2
Vapor vent systems are provided in pressure carburetors to eliminate fuel vapor created by the fuel pump, heat in the engine compartment, and the pressure drop across the poppet valve. If the vapor vent valve sticks open or the float becomes filled with fuel and sinks, a continuous flow of fuel and vapor occurs through the vent line. The vent line feeds back to the fuel tank.
REFERENCE: AC65-12A, Page 126

4659. ANSWER #1
Idle speed on engines using a float-type carburetor is adjusted by limiting how far the throttle valve will close. This is usually done with an adjustable throttle stop or linkage.
REFERENCE: ITP POWERPLANT, Chapter 6, Page 26
              AC65-12A, Page 146

4660. ANSWER #2
The primary reason for trimming a turbine engine is to ensure that the desired thrust is obtained when the power lever is moved to a specific position.
REFERENCE: ITP POWERPLANT, Chapter 6, Page 61
              AC65-12A, Page 167

4661. ANSWER #3
The most popular and most often used fuel control has been the hydromechanical. The trend now, however, is to incorporate more and more electronics into the fuel metering system. Many engines are now using a hydromechanical control with an electronic system added to it. The electronic part of the system fine tunes the fuel metering to improve the efficiency and economy.
REFERENCE: AC65-12A, Page 149

4662. ANSWER #4
The ideal conditions for trimming a turbine engine are no wind, low humidity, and standard day temperature and pressure. Because standard day conditions seldom exist, the trim charts compensate for the conditions as they exist. If there is a little bit of wind, the airplane should be headed into the wind. In extremely windy conditions, trimming should not be attempted.
REFERENCE: AC65-12A, Page 169

4663. ANSWER #4
Figure 3 in the question book, which comes from page 110 in the AC65-12A, shows the fuel/air ratio that exists from idle to rated power. The mixture used at rated power is richer than that used in cruise, but leaner than what is used at idle. The rich mixtures used at idle and at rated power help to keep the engine cool.
REFERENCE: AC65-12A, Pages 110 and 113 (Figure 3-1)

4664. ANSWER #3
The degree of atomization or vaporization is the extent to which fine spray is produced; the more fully the mixture is reduced to fine spray and vaporized, the greater is the efficiency of the combustion process. The air bleed in the main discharge nozzle passage aids in the atomization and vaporization of the fuel. If the fuel is not fully vaporized, the mixture may run "lean" even though there is an abundance of fuel present.
REFERENCE: AIRCRAFT POWERPLANTS, Page 78

4665. ANSWER #4
On all aircraft installations where manifold pressure gages are used, the manifold pressure gage will give a more consistent and larger indication of power change at idle speed than will the tachometer. Therefore, utilize the manifold pressure gage when adjusting the idle fuel/air mixture.
REFERENCE: ITP POWERPLANT, Chapter 6, Page 26
             AC65-12A, Page 144

4666. ANSWER #3
The idle speed and idle mixture on an engine is known to be properly set when the RPM is at its maximum and the manifold pressure is at its minimum. When the mixture control is pulled back toward the idle cutoff position, there should be a slight rise in manifold pressure just before the engine dies. This indicates that the mixture is rich, as it should be at idle, and it has passed through the best power ratio on its way to idle cutoff.
REFERENCE: ITP POWERPLANT, Chapter 6, Page 26
             AC65-12A, Page 144

4667. ANSWER #1
The opening of the throttle valve is what determines the amount of air which will flow through the carburetor. When the engine is being rotated in an attempt to start, a high vacuum will be created on the engine side of the throttle. If the throttle valve opening is less than normal, the high vacuum will act on the idle jet and draw more fuel in relation to air than it should. This will cause a richer than normal mixture during starting.
REFERENCE: AIRCRAFT POWERPLANTS, Pages 80 and 81

4668. ANSWER #4
On installations that do not use a manifold pressure gage, it will be necessary to observe the tachometer for an indication of correct idle mixture. With most installations, the idle mixture should be adjusted to provide an RPM rise prior to decreasing as the engine ceases to fire. This RPM increase will vary from 10 to 50 RPM, depending on the installation.
REFERENCE: ITP POWERPLANT, Chapter 6, Page 26
             AC65-12A, Pages 144 and 145

4669. ANSWER #3
When a new carburetor is installed on an engine, the idle speed and mixture must be set. These adjustments should not be made until the engine has been operating long enough to have normal cylinder head temperatures.
REFERENCE: ITP POWERPLANT, Chapter 6, Page 26
             AC65-12A, Page 144

4670. ANSWER #2
The back-suction type mixture control system controls the mixture by allowing a certain amount of the venturi low pressure to act on the fuel in the float chamber. The low pressure acting on the fuel in the float chamber reduces the pressure differential that exists between the fuel and the main discharge nozzle. There is also an atmospheric line, with an adjustable valve in it, opening into the float chamber. By varying the amount of atmospheric pressure allowed to enter the chamber, which affects the pressure differential between the fuel and the discharge nozzle, the mixture to the engine is controlled.
REFERENCE: ITP POWERPLANT, Chapter 6, Page 17
             AC65-12A, Page 121

4671. ANSWER #2
Water vapor is a non-combustible gas. When water vapor is in the air, it reduces the volumetric efficiency of the engine, and therefore reduces the power of the engine. This holds true with recip engines, but not with turbine.
REFERENCE: ITP POWERPLANT, Chapter 6, Page 4
             AC65-9A, Page 239

4672. ANSWER #2

At low RPM, namely idle speeds, there is insufficient airflow through the carburetor for the main discharge nozzle to operate properly. For this reason, a separate idle jet is used near the edge of the throttle valve to supply fuel for idling. If the idle jet becomes clogged, the engine will not idle.
REFERENCE: ITP POWERPLANT, Chapter 6, Page 18
AC65-12A, Page 120

4673. ANSWER #1

Aircraft engines using pressure-type carburetors are generally started using the primer, with the mixture control positioned in idle cutoff. As soon as the engine starts, the mixture control is moved to full rich.
REFERENCE: AC65-9A, Page 490
AIRCRAFT POWERPLANTS, Page 125

4674. ANSWER #2

Operating a recip engine with water injection allows the mixture to be leaned out to the best power setting without fear of detonation because of the cooling effect of the water.
REFERENCE: AC65-12A, Page 147

4675. ANSWER #1

The amount of pressure drop at the venturi of a carburetor depends on the velocity of the intake air. The higher the velocity, the greater the pressure drop.
REFERENCE: ITP POWERPLANT, Chapter 6, Page 14
AC65-12A, Page 113

4676. ANSWER #1

When an engine is operated on a lean mixture, the cylinder head temperature gage should be watched closely. At high power settings, a lean mixture will cause extremely high cylinder head temperatures. If the mixture is excessively lean, the engine may backfire through the induction system or stop completely.
REFERENCE: ITP POWERPLANT, Chapter 6, Page 6
AC65-12A, Page 113

4677. ANSWER #3

Air with water vapor in it weighs less than dry air, because the gas water vapor weighs less than air.
REFERENCE: ITP POWERPLANT, Chapter 6, Page 4
AC65-9A, Page 239

4678. ANSWER #1

When looking at this question, it is necessary to realize that the ratio was written the same way the AC65-12A shows it. If the ratio is taken literally, answers 2 and 3 are correct. A fuel/air mixture of 11:1 would be 11 parts fuel and 1 part air, and it would be too rich to burn. Looking at it realistically, it is probably meant to be 11 parts air to 1 part fuel.
REFERENCE: AC65-12A, Page 112

4679. ANSWER #4

An economizer system is designed to enrich the fuel mixture when the throttle is advanced to full power. The economizer allows the mixture to be adjusted to a lean setting for most engine operations, because the mixture will be automatically enriched at full power.
REFERENCE: ITP POWERPLANT, Chapter 6, Page 20
AC65-12A, Pages 121 and 122

4680. ANSWER #4

When the throttle valve is opened quickly, a large volume of air rushes through the air passage of the carburetor. The normal fuel metering system of the carburetor cannot respond quickly enough to match the airflow, so an accelerator pump is provided to add additional fuel during rapid movements of the throttle.
REFERENCE: ITP POWERPLANT, Chapter 6, Pages 19 and 20
AC65-12A, Page 121

4681. ANSWER #2

A pressurizing and dump valve is used on a turbine engine which has dual-line duplex fuel nozzles. The pressurizing part of the valve separates the primary and secondary fuel flows to the fuel nozzles. During starting, the pressurizing valve is closed, allowing only primary fuel to flow to the nozzle. The dump part of the valve drains or "dumps" the fuel manifold on the engine shutdown. When the engine is being started or is running, the dump valve is closed.
REFERENCE: ITP POWERPLANT, Chapter 6, Pages 79 and 80
AC65-12A, Page 173

4682. ANSWER #4
Of the air consumed by a turbine engine, only a small part is used to support combustion (approximately 25%). For this reason, water vapor in the air has very little effect on the thrust produced by a jet engine.
REFERENCE: AC65-12A, Page 489

4683. ANSWER #2
A pressurizing and dump valve is used on a turbine engine which has dual-line duplex fuel nozzles. The pressurizing part of the valve separates the primary and secondary fuel flows to the fuel nozzle. When the engine is shut down, the pressurizing valve is closed, and will stay closed until the engine is running and at a medium power setting. When the pressurizing valve opens, it allows secondary fuel to flow along with the primary fuel. The dump part of the valve drains or "dumps" the fuel manifold on engine shutdown.
REFERENCE: ITP POWERPLANT, Chapter 6, Pages 79 and 80
            AC65-12A, Page 173

4684. ANSWER #2
The purpose of an automatic mixture control is to decrease the fuel flow from the carburetor as the aircraft climbs to higher altitudes. Because of the decreased air density at altitude, the fuel flow needs to be reduced to avoid operating with an excessively rich mixture. If the automatic mixture control were to malfunction and stick in a high altitude position, the engine would run extremely lean at lower altitudes, and this would cause high cylinder head temperatures when operating at high power.
REFERENCE: AC65-12A, Page 131
            AIRCRAFT POWERPLANTS, Page 95

4685. ANSWER #4
In modern day turbine engine fuel controls, there are many different parameters monitored. The ones listed in the AC65-12A include power lever position, engine RPM, compressor inlet pressure or temperature, and burner pressure or compressor discharge pressure.
REFERENCE: ITP POWERPLANT, Chapter 6, Page 51
            AC65-12A, Page 149

4686. ANSWER #1
In one model of fuel control used on the Pratt and Whitney JT3-C turbojet engine, there is a valve called a surge and temperature limiting valve. This valve overrides the action of the speed governor during rapid acceleration to ensure that surge and temperature operational limits of the engine are not exceeded. Surge is another way of describing a compressor stall.
REFERENCE: AC65-12A, Page 165

4687. ANSWER #1
Detonation is the spontaneous combustion of the unburned fuel/air mixture, ahead of the flame front, after ignition of the charge. The explosive burning during detonation results in an extremely rapid pressure rise and high instantaneous temperature.
REFERENCE: ITP POWERPLANT, Chapter 6, Page 5
            AC65-12A, Page 444

4688. ANSWER #1
When a carburetor is found to have fuel leaking from the discharge nozzle, the fuel level in the float chamber has to be suspect. This problem is caused by the fuel level being too high in the chamber. The high fuel level can be caused by an improperly adjusted float, a float which has a hole in it, dirt trapped between the needle and seat, or a worn needle and seat.
REFERENCE: ITP POWERPLANT, Chapter 6, Page 15
            AC65-12A, Page 117
            AIRCRAFT POWERPLANTS, Page 95

4689. ANSWER #3
When an aircraft's boost pump pressure is greater than that of the pressure pump, a bypass valve in the pressure pump opens and allows fuel to flow around it and go out to the engine. This would happen if the pressure pump was inoperative.
REFERENCE: ITP AIRFRAME, Chapter 9, Page 24
            AC65-9A, Page 87

4690. ANSWER #4
Centrifugal-type boost pumps are not classified as positive displacement pumps. Once their pressure builds to approximately 20 to 30 psi, fuel starts slipping around the impeller rather than being pumped out to the system.
REFERENCE: ITP AIRFRAME, Chapter 9, Page 22
            AC65-9A, Page 85

4691. ANSWER #1
According to the Federal Aviation Regulations, the engine fuel shutoff valve may not be on the engine side of the firewall.
REFERENCE: FAR 23.995 (b)

4692. ANSWER #4
Boost pumps in an aircraft fuel system serve the following purposes:
1. They allow fuel to be transferred from tank to tank within the aircraft.
2. They help prevent vapor lock by separating the air from fuel.
3. They provide a positive flow of fuel to the engine driven pump.
4. They serve as emergency pumps in case the engine driven pump fails.
REFERENCE: AC65-9A, Page 85

4693. ANSWER #2
An ejector helps cause fluid movement by creating a low pressure area at the point where the movement is needed. The fuel transfer ejectors in this system help transfer the fuel from the main tank to the boost pump sump.
REFERENCE: AIRCRAFT MAINTENANCE AND REPAIR, Page 401

4694. ANSWER #2
When an aircraft's boost pump pressure is greater than that of the pressure pump, a bypass valve in the pressure pump opens and allows fuel to flow around it and go out to the engine. This would happen if the pressure pump was inoperative.
REFERENCE: ITP AIRFRAME, Chapter 9, Page 24
AC65-9A, Page 87

4695. ANSWER #1
The purpose of the engine-driven fuel pump is to deliver a continuous supply of fuel at the proper pressure at all times during engine operation. One type of pump which is widely used is the positive-displacement, rotary-vane pump.
REFERENCE: AC65-9A, Page 86

4696. ANSWER #4
One of the many purposes of a boost pump is to supply fuel under pressure for priming when starting the engine.
REFERENCE: AC65-9A, Page 85

4697. ANSWER #2
Since the engine-driven fuel pump normally discharges more fuel than the engine requires, there must be some way of relieving excess fuel to prevent excessive fuel pressures at the fuel inlet of the carburetor. This is accomplished through the use of a spring loaded relief valve that bypasses excess fuel back to the inlet side of the pump.
REFERENCE: ITP AIRFRAME, Chapter 9, Page 24
AC65-9A, Page 86

4698. ANSWER #4
The purpose of the engine-driven fuel pump is to deliver a continuous supply of fuel at the proper pressure at all times during engine operation. One type of pump which is widely used is the positive-displacement, rotary-vane pump.
REFERENCE: AC65-9A, Page 86

4699. ANSWER #4
In a compensated vane-type fuel pump, the pressure of fuel delivered to the inlet of the carburetor varies according to the altitude and the corresponding atmospheric pressure. It is varied by allowing a set spring tension and either atmospheric or carburetor inlet air pressure on a diaphragm to act on the pump's relief valve. This determines the pressure at which the relief valve will bypass fuel back to the pump's inlet.
REFERENCE: AC65-9A, Pages 86 and 87

4700. ANSWER #3
The two types of fuel nozzles used in turbine engine combustion chambers are the simplex and the duplex. The duplex is the most common.
REFERENCE: ITP POWERPLANT, Chapter 6, Page 78
AC65-12A, Page 172

4701. ANSWER #1
The fuel line between the engine fuel pump and the carburetor is not susceptible to vapor lock. By the time the fuel gets to this point, it has probably gone through a boost pump, which removes the vapor from the fuel.
REFERENCE: AC65-12A, Page 109

4702. ANSWER #3
According to the Federal Aviation Regulations, aircraft that are certified in a standard manner, such as normal category airplanes, must have a positive means of shutting off the fuel to all engines.
REFERENCE: FAR 23.1189 (a) (1)

4703. ANSWER #3
The main fuel strainer in an aircraft's fuel system is installed so that the fuel flows through it before reaching the engine-driven pump. It should be located at the lowest point in the fuel system so that water and other debris will collect in it and can be readily drained.
REFERENCE: AC65-9A, Page 84

4704. ANSWER #1
Every effort should be made to avoid a potential fire hazard by physically separating electric wiring and flammable fluid carrying lines. When separation is impractical, the electric wire should be located above the flammable fluid line and securely clamped to the structure.
REFERENCE: AC43.13-1A, Page 187

4705. ANSWER #2
One nice feature of having a centrifugal boost pump in a fuel system is its ability to separate air and vapor from the fuel it is pumping. This helps reduce the possibility of vapor lock.
REFERENCE: ITP AIRFRAME, Chapter 9, Page 22
AC65-9A, Page 85

4706. ANSWER #4
According to FAR 23, the fuel flow rate of a gravity feed system must be at least 150% of the takeoff fuel consumption of the engine.
REFERENCE: FAR 23.955 (b)

4707. ANSWER #1
One of the functions of fuel boost pumps is to supply a positive flow of fuel to the engine driven fuel pump.
REFERENCE: ITP AIRFRAME, Chapter 9, Page 22
AC65-9A, Page 85

4708. ANSWER #2
The purpose of a fuel pump relief valve is to prevent excessive pressures from reaching the inlet of the carburetor. In a normally operating system, the fuel pump is always putting out more fuel than the engine needs. The fuel pump, therefore, is always bypassing fuel back to the pump inlet. If the relief valve is sticking when the engine speeds up and the pump puts out more fuel, the fuel pressure will increase because the relief valve won't be able to react to the increased fuel flow.
REFERENCE: ITP AIRFRAME, Chapter 9, Page 24
AC65-9A, Page 86

4709. ANSWER #3
According to FAR 23, each fuel tank outlet must have a fuel strainer.
REFERENCE: FAR 23.977 (a)

4710. ANSWER #4
Fuel pump relief valves designed to compensate for atmospheric pressure variations are known as balanced-type relief valves. See the explanation to question #3699 for an explanation of the valve's operation.
REFERENCE: AC65-9A, Pages 86 and 87

4711. ANSWER #2
To reduce the possibility of vapor lock, fuel lines should be kept away from sources of heat. In addition, sharp bends and steep rises should be avoided.
REFERENCE: AC65-12A, Page 109

4712. ANSWER #4
In order to maintain the proper balance and stability in a large airplane, it is essential that the fuel system be capable of moving fuel from one tank to another. For this reason, aircraft fuel systems with multiple tanks have a crossfeed system that allows fuel to be moved from one tank to another.
REFERENCE: AC65-9A, Page 93

4713. ANSWER #4
A float-type carburetor operated with an engine driven fuel pump is normally supplied with an inlet pressure of 3 to 6 psi. At pressures higher than this, there might be a problem with the float needle and seat leaking.
REFERENCE: AIRCRAFT PROPULSION POWERPLANTS, Page 421

4714. ANSWER #3
Because engine driven fuel pumps supply more fuel than the engine can consume, the pump must be fitted with a relief valve to bypass excess fuel back to the pump inlet.
REFERENCE: AC65-9A, Page 86

4715. ANSWER #1
A rotary-vane pump, such as the ones used as fuel pumps, are positive displacement. This means that for each revolution of the pump, a fixed quantity of fluid is deposited in the outlet port.
REFERENCE: AC65-9A, Page 86

4716. ANSWER #2
The relief valve used in the engine driven fuel pump is adjustable, so the pressure delivered to the engine can be fine tuned to meet the particular needs.
REFERENCE: AC65-9A, Page 86

4717. ANSWER #3
This question is the same as #4695 and #4698. The vane-type pump is often used as an engine driven fuel pump.
REFERENCE: AC65-9A, Page 86

4718. ANSWER #2
Although gasoline has more heat energy per pound than kerosene (20,000 BTU's compared to 18,500 BTU's), kerosene weighs more than gasoline and therefore has more heat energy per gallon.
REFERENCE: ITP POWERPLANT, Chapter 6, Pages 10 and 49

4719. ANSWER #2
The principle advantages of the duplex fuel nozzle over the simplex fuel nozzle are the improved atomization and more uniform flow pattern it offers.
REFERENCE: AC65-12A, Page 172

4720. ANSWER #1
With an aircraft engine, it must be possible to increase and decrease power at will to obtain the power required for any operating condition. In a turbine engine, the fuel control is what allows the power changes to occur. With a turbine engine, however, the changes cannot be allowed to occur too quickly. Too rapid an acceleration or deceleration of the engine could cause a compressor stall, which could lead to a rich "blowout" or lean "dieout". The turbine engine fuel control has built-in features which try to prevent the compressor stall from occurring.
REFERENCE: ITP POWERPLANT, Chapter 6, Page 49
            AC65-12A, Page 165

4721. ANSWER #2
The purpose of trimming a turbine engine is to adjust the idle RPM and maximum speed, or the idle RPM and target EPR, depending on the type of engine.
REFERENCE: ITP POWERPLANT, Chapter 6, Page 61
            AC65-12A, Page 167

4722. ANSWER #1
A micron is a unit of measurement equal to 1/25,000 of an inch. Wafer screen filters carry a micron rating, and they can be built with a great filtering capability. The AC65-12A, however, classifies micron as a type of filter, and says it has the greatest filtering action.
REFERENCE: AC65-12A, Page 170

4723. ANSWER #3
The flow divider in a duplex fuel nozzle creates two separate fuel supplies, called primary and secondary. The primary fuel flows all the time, from starting all the way to takeoff thrust. The secondary fuel only flows when the pressure has built up enough to unseat the flow divider. This normally happens in the mid RPM range.
REFERENCE: ITP POWERPLANT, Chapter 6, Page 78
            AC65-12A, Page 172

4724. ANSWER #1
The flow divider in a turbine engine duplex fuel nozzle opens because of fuel pressure.
REFERENCE: ITP POWERPLANT, Chapter 6, Page 78
            AC65-12A, Page 172

4725. ANSWER #1
Induction system ice in a recip engine can be prevented or eliminated by raising the temperature of the air that passes through the system, using a preheater located upstream near the induction system inlet and well ahead of the dangerous icing zones.
REFERENCE: ITP POWERPLANT, Chapter 7, Page 3
            AC65-12A, Pages 71 and 72

4726. ANSWER #4
See the explanation to #4725.
REFERENCE: ITP POWERPLANT, Chapter 7, Page 3
            AC65-12A, Pages 71 and 72

4727. ANSWER #1
Icing conditions with a normally aspirated engine, using a float-type carburetor, are most severe in the temperature range of 30 degrees to 40 degrees Fahrenheit. Even though water does not freeze at 40 degrees Fahrenheit, the cooling effect of the fuel vaporizing can easily bring the temperature down to freezing.
REFERENCE: AC65-9A, Page 74

4728. ANSWER #2
Some of the large reciprocating engines have supplemented the carburetor heat system with an alcohol deicing system. The system allows the pilot to spray alcohol into the inlet of the carburetor to remove ice and to assist the warm air in keeping the carburetor free from ice.
REFERENCE: ITP POWERPLANT, Chapter 7, Page 4

4729. ANSWER #3
Icing conditions in an engine's induction system can be detected by a reduction in engine power when the throttle position remains the same. If the aircraft is equipped with a fixed-pitch propeller, the engine RPM will decrease. With a constant-speed propeller, the manifold pressure will decrease and the engine power will drop, even though the engine RPM remains constant.
REFERENCE: AIRCRAFT POWERPLANTS, Page 138

4730. ANSWER #4
Because of the cooling effect of fuel vaporizing and the low pressures that exist, the throat of a carburetor will form ice before any other part of the aircraft.
REFERENCE: ITP POWERPLANT, Chapter 7, Page 1
            AC65-12A, Page 73

4731. ANSWER #1
Heating the air in the inlet duct and spraying alcohol in the inlet to the carburetor are two methods of eliminating carburetor icing.
REFERENCE: ITP POWERPLANT, Chapter 7, Page 4
            AC65-12A, Pages 72 and 73

4732. ANSWER #3
Because fuel injection systems generally inject the fuel directly into the cylinder, there is not a prime location for ice to develop like there is in the engine using a float-type carburetor. For this reason, fuel injected engines seldom use a carburetor air heater.
REFERENCE: ITP POWERPLANT, Chapter 7, Page 3
            AIRCRAFT POWERPLANTS, Page 114

4733. ANSWER #2
Icing conditions in an engine with a constant speed propeller will cause the manifold pressure to decrease, while the engine RPM remains constant. If carburetor heat is applied to this engine, with an icing condition, the manifold pressure will rise as the ice melts.
REFERENCE: AIRCRAFT POWERPLANTS, Page 138

4734. ANSWER #4
In the induction system of an unsupercharged engine, the pressure is always less than atmospheric when the engine is running. The induction system includes everything from the carburetor inlet to the intake valve. A higher pressure would exist in the carburetor air scoop than at any point in the induction system.
REFERENCE: AIRCRAFT POWERPLANTS, Page 135

4735. ANSWER #3
When carburetor heat is used, the increase in air temperature causes the air to expand and decrease in density. This action reduces the weight of the charge delivered to the cylinders and causes a loss in power because of decreased volumetric efficiency. In addition, high intake air temperature may cause detonation and engine failure, especially during takeoff and high-power operation.
REFERENCE: AC65-12A, Page 72

4736. ANSWER #1
During the check of a two-speed single-stage supercharger, a normal indication when shifting from low to high speed is an increase in engine RPM, an increase in manifold pressure, and a momentary drop in oil pressure. The increase in RPM and manifold pressure are caused by the additional engine power available in the high supercharger mode. The momentary decrease in oil pressure is caused by the large amount of oil required to shift the supercharger into high speed, which momentarily decreases the flow to the rest of the engine.
REFERENCE: AC65-12A, Page 441

4737. ANSWER #1
On large volume engines ranging from 450 H.P. upwards, in which the volume of mixture is to be handled at higher velocities and turbulence is a more important factor, either a vane or airfoil type diffuser is widely used. The vanes or airfoil sections straighten the airflow within the diffuser chamber to obtain an efficient flow of gases.
REFERENCE: AC65-12A, Page 77

4738. ANSWER #3

The pressure of a gas and its density are directly proportional to each other. As the manifold pressure of an engine increases, the density of the charge going to the cylinders increases proportionally.
REFERENCE: AC65-12A, Page 76

4739. ANSWER #4

The volumetric efficiency of an engine is a comparison of the volume of fuel/air charge inducted into the cylinders to the total piston displacement of the engine. An engine which draws in less volume than it has displacement, has a volumetric efficiency of less than 100%. An engine equipped with a high-speed internal or external blower may have a volumetric efficiency greater than 100%.
REFERENCE: AC65-12A, Page 38

4740. ANSWER #3

The purpose of the waste gate in a turbocharger system is to route some of the exhaust gases through the turbocharger, and to allow the remainder of the exhaust to flow out through the exhaust system. The amount of engine boost is controlled by not allowing too much exhaust to flow through the turbocharger. If the waste gate were to be closed with the engine operating, serious damage could be done to the engine because of overboost.
REFERENCE: AC43.13-1A, Page 290
                        AIRCRAFT POWERPLANTS, Page 155

4741. ANSWER #1

On large volume engines ranging from 450 H.P. upwards, in which the volume of mixture is to be handled at higher velocities and turbulence is a more important factor, either a vane or airfoil type diffuser is widely used. The vanes or airfoil sections straighten the airflow within the diffuser chamber to obtain an efficient flow of gases. In addition to straightening the airflow, the vanes diffuse the air, which means they increase its pressure.
REFERENCE: AC65-12A, Page 77

4742. ANSWER #2

When carburetor heat is used, the increase in air temperature causes the air to expand and decrease in density. This action reduces the weight of the charge delivered to the cylinders and causes a loss in power because of decreased volumetric efficiency. In addition, high intake air temperature may cause detonation and engine failure, especially during takeoff and high-power operation. Even if the cockpit controls indicate that carburetor heat is off, the problem could still exist if the linkage on the heat valve is improperly adjusted.
REFERENCE: AC65-12A, Page 72

4743. ANSWER #4

The purpose of the waste gate in a turbocharger system is to direct the gases through the turbine wheel, or to allow them to go out the tail pipe, or a combination of both. If the waste gate is closed, all the exhaust gases must pass through the turbine wheel before they go out the tail pipe.
REFERENCE: ITP POWERPLANT, Chapter 7, Page 8
                        AC65-12A, Page 86

4744. ANSWER #4

To really be considered a supercharger, the system must be able to raise the manifold pressure above 30 inches of mercury.
REFERENCE: AIRCRAFT POWERPLANTS, Page 141

4745. ANSWER #3

The density controller in a turbocharger system is designed to limit the manifold pressure below the turbocharger's critical altitude.
REFERENCE: AC65-12A, Page 86

4746. ANSWER #3

The rate-of-change controller in a turbocharger system controls the rate at which the turbocharger compressor discharge pressure will increase.
REFERENCE: AC65-12A, Page 86

4747. ANSWER #3

The ultimate control on the speed of the turbine wheel in a turbocharger is the energy available in the exhaust gases and the ability of the turbine wheel to use that energy. The component which controls how much of the exhaust gas is allowed to flow through the turbine is the waste gate. The waste gate position, of course, is controlled by other components in the system.
REFERENCE: ITP POWERPLANT, Chapter 7, Page 9
                        AC65-12A, Page 86

4748. ANSWER #3

As an aircraft gains altitude, the decrease in air density tries to make its engine less powerful. If the density of the charge entering the engine were to be maintained, the engine would be able to maintain its power output. Turbocharging offers the engine a way of doing this. Turbochargers compress the air before it enters the engine, thereby maintaining the air density (manifold pressure) up to the critical altitude of the engine.
REFERENCE: ITP POWERPLANT, Chapter 7, Page 5
                        AC65-12A, Page 84

4749. ANSWER #2
A sea level boosted turbocharger system is automatically regulated by three components. They are: (1) The exhaust bypass valve assembly, (2) the density controller, and (3) the differential pressure controller.
REFERENCE: AC65-12A, Page 86

4750. ANSWER #1
The differential pressure controller reduces the unstable condition known as "bootstrapping" during part-throttle operation. Bootstrapping is an indication of unregulated power change that results in the continual drift of manifold pressure.
REFERENCE: AC65-12A, Page 89

4751. ANSWER #1
A pressure ratio controller in a turbocharging system controls the waste-gate actuator above critical altitude.
REFERENCE: AC65-12A, Page 84

4752. ANSWER #3
An externally driven supercharger is a turbocharger. Turbochargers are driven by engine exhaust gases flowing through a turbine wheel.
REFERENCE: ITP POWERPLANT, Chapter 7, Page 5
               AC65-12A, Page 84

4753. ANSWER #2
The typical turbosupercharger for a large reciprocating engine is composed of three main parts:
   1. The compressor assembly.
   2. The exhaust gas turbine assembly.
   3. The pump and bearing casing.
REFERENCE: AC65-12A, Page 81

4754. ANSWER #1
When carburetor heat is turned on, the increase in air temperature brings with it a decrease in air density. Because the carburetor meters fuel in relation to the volume and velocity through the venturi, the same amount of fuel will be mixed with the heated air. This will produce a rich mixture.
REFERENCE: AIRCRAFT POWERPLANTS, Page 92

4755. ANSWER #2
A carburetor air heater should be placed in the cold position when starting an engine. This is done to prevent damage to the heater valves in case the engine should backfire.
REFERENCE: AC65-12A, Page 72

4756. ANSWER #1
Because heated air is less dense than cool air, the weight of the fuel/air charge is decreased when carburetor heat is used.
REFERENCE: AC65-12A, Page 72
               AIRCRAFT POWERPLANTS, Page 92

4757. ANSWER #4
When carburetor heat is turned on, the increase in air temperature brings with it a decrease in air density. Because the carburetor meters fuel in relation to the volume and velocity through the venturi, the same amount of fuel will be mixed with the heated air. This will produce a rich mixture.
REFERENCE: AIRCRAFT POWERPLANTS, Page 92

4758. ANSWER #3
When carburetor heat is turned on, the increase in air temperature brings with it a decrease in air density. Because the carburetor meters fuel in relation to the volume and velocity through the venturi, the same amount of fuel will be mixed with the heated air. This will produce a rich mixture. If the carburetor is fitted with an automatic mixture control (AMC), the enriching of the mixture will be corrected once the AMC takes over.
REFERENCE: AIRCRAFT POWERPLANTS, Page 92
               AC65-12A, Page 131

4759. ANSWER #4
Electric priming valves would normally receive their fuel from the fuel boost pumps. There is one type of valve, described in the text "Aircraft Powerplants", which does use fuel from the engine-driven pump as an alternate fuel supply. The carburetor is normally used as a source of fuel for a hand-operated primer pump.
REFERENCE: AIRCRAFT POWERPLANTS, Page 129

4760. ANSWER #3
Because of the possibility of liquid lock, the priming system on a radial engine only primes the cylinders from the horizontal up. On a nine cylinder radial engine, this would be cylinders one, two, three, eight, and nine.
REFERENCE: AIRCRAFT PROPULSION POWERPLANTS, Page 158

4761. ANSWER #2
If a fire breaks out in the induction system when a recip engine is being started, one method of extinguishing the fire is to continue to run the engine. By doing this, the fire will be drawn into the engine and the problem should go away. If the fire does not go out, however, the engine should be shut down and a CO_2 fire extinguisher used to put out the fire.
REFERENCE: AC65-9A, Page 491

4762. ANSWER #2
This question is asking the same thing as #4761. If a fire breaks out in the induction system when a recip engine is being started, one method of extinguishing the fire is to continue to run the engine. By doing this, the fire will be drawn into the engine and the problem should go away. If the fire does not go out, however, the engine should be shut down and a CO_2 fire extinguisher used to put out the fire.
REFERENCE: AC65-9A, Page 491

4763. ANSWER #2
To aid the vaporization of the fuel, many horizontally opposed engines have the carburetor mounted on the oil sump and the induction pipes pass through the oil. This serves the double function of cooling the oil and at the same time warming the induction air without the danger of heating it to the point that detonation could occur.
REFERENCE: ITP POWERPLANT, Chapter 7, Page 4

4764. ANSWER #4
A typical air scoop is simply an opening facing into the airstream. This scoop receives ram air, usually augmented by propeller slipstream. The effect of the air velocity is to "supercharge" the air a small amount, thus adding to the total weight of air received by the engine.
REFERENCE: AC65-12A, Page 71
            AIRCRAFT POWERPLANTS, Page 135

4765. ANSWER #1
When carburetor heat is used, the increased temperature of the air causes it to expand and decrease in density. This reduces the power of the engine because of the reduced volumetric efficiency. In addition, the high intake air temperatures may cause detonation, especially during takeoff and high-power operation.
REFERENCE: AC65-12A, Page 72

4766. ANSWER #1
Cowling and baffles are designed to force air over the cylinder cooling fins. The baffles direct the air close around the cylinders and prevent it from forming hot pools of stagnant air while the main streams rush by unused.
REFERENCE: ITP POWERPLANT, Chapter 7, Pages 27 and 28
            AC65-12A, Pages 314 and 315

4767. ANSWER #2
Some aircraft use augmentors to provide additional cooling airflow for the engine. The velocity of the exhaust gases leaving the augmentors creates a low pressure area which helps draw cooling air in.
REFERENCE: ITP POWERPLANT, Chapter 7, Page 29
            AC65-12A, Page 315

4768. ANSWER #4
Cylinder deflectors and baffles are designed to force air over the cylinder cooling fins. The baffles direct the air close around the cylinders and prevent it from forming hot pools of stagnant air while the main streams rush by unused. Because of the importance of the cylinder baffles, they should be inspected regularly and repaired if any damage is noticed. Even a small amount of damage could cause a localized hot spot and an eventual malfunctioning engine.
REFERENCE: ITP POWERPLANT, Chapter 7, Pages 27 and 28
            AC65-12A, Pages 316, 317 and 320

4769. ANSWER #1
This question is the same as #4766. Cylinder deflectors and baffles are designed to force air over the cylinder cooling fins. The baffles direct the air close around the cylinders and prevent it from forming hot pools of stagnant air while the main streams rush by unused.
REFERENCE: ITP POWERPLANT, Chapter 7 , Pages 27 and 28
            AC65-12A, Pages 314 and 315

4770. ANSWER #2
Cracks in the cooling fins of a cylinder are allowed, providing they are within the manufacturer's allowable limits. Stop drilling a crack is a possibility, but not in the cylinder head. Removing the damaged area and then contour filing is a common practice, but again it must be done within the manufacturer's allowable limits.
REFERENCE: AC65-12A, Page 319

4771. ANSWER #2
When performing repairs to the cooling fins of a cylinder, the engine manufacturer's service or overhaul manual should be consulted to ensure the repair is within limits.
REFERENCE: AC65-12A, Page 319

4772. ANSWER #3
If a cooling fin is found bent on an aluminum cylinder head, it should be left alone if it is not cracked. Aluminum cooling fins are very brittle, and any attempt to straighten it could cause it to crack or break.
REFERENCE: AIRCRAFT POWERPLANTS, Page 294

4773. ANSWER #4
On an air-cooled reciprocating engine, cooling fins are located on the cylinder head, cylinder barrel, and on the inside of the piston head.
REFERENCE: ITP POWERPLANT, Chapter 1, Pages 27, 29 and 32
                      AC65-12A, Pages 15 and 18

4774. ANSWER #3
Cowling and baffles are designed to force air over the cylinder cooling fins. The baffles direct the air close around the cylinders and prevent it from forming hot pools of stagnant air while the main streams rush by unused.
REFERENCE: ITP POWERPLANT, Chapter 7, Pages 27 and 28
                      AC65-12A, Pages 314 and 315

4775. ANSWER #2
The most common means of controlling recip engine cooling is the use of cowl flaps. When extended for increased cooling, the cowl flaps produce drag and sacrifice streamlining for the added cooling. On takeoff, the cowl flaps are opened only enough to keep the engine below the red-line temperature. During ground operations, the cowl flaps should be opened wide since drag does not matter.
REFERENCE: AC65-12A, Page 315

4776. ANSWER #1
Virtually all turbine engines in use today make use of oil coolers. Most of these engines use dry-sump oil systems, and many use some means of air cooling the turbine section and bearings.
REFERENCE: AC65-12A, Page 302

4777. ANSWER #1
Any time an electrical component fails to operate, the first thing to check is power to the component. This means checking the circuit breaker or fuse that protects that circuit.
REFERENCE: AIRCRAFT ELECTRICITY AND ELECTRONICS, Page 168

4778. ANSWER #3
Some aircraft use augmentors to provide additional cooling airflow for the engine. The velocity of the exhaust gases leaving the augmentors creates a low pressure area which helps draw cooling air in. In addition to the cooling of the engine, augmentor systems can also be used to provide heated air for cabin heat, anti-icing and deicing systems.
REFERENCE: ITP POWERPLANT, Chapter 7, Page 29
                      AC65-12A, Page 315

4779. ANSWER #1
The amount of damaged fin area on a cylinder is a determining factor in whether or not the cylinder can remain in service. The reason for removal if there is too much fin damage in a localized area is the possibility of a hot spot developing.
REFERENCE: AC65-12A, Page 319

4780. ANSWER #4
Because of the difference in temperature in the various sections of the cylinder head, it is necessary to provide more cooling-fin area on some sections than on others. The exhaust valve region is the hottest part of the internal surface; therefore, more fin area is provided around the outside of the cylinder in this section.
REFERENCE: ITP POWERPLANT, Chapter 1, Page 29
                      AC65-12A, Page 18

4781. ANSWER #3
In the operation of a recip powered helicopter, the ram air pressure from the rotor system is usually not sufficient to cool the engine, particularly when the craft is hovering. For this reason, a large engine-driven fan is installed in a position to maintain a strong flow of air across and around the cylinders and other parts of the engine.
REFERENCE: AIRCRAFT POWERPLANTS, Page 33

4782. ANSWER #3
In a typical aircraft reciprocating engine, half of the heat generated in the engine goes out with the exhaust, and the other half is absorbed by the engine. Circulating oil picks up part of this soaked-in heat and transfers it to the airstream through the oil cooler. The engine cooling system takes care of the rest.
REFERENCE: AC65-12A, Page 314

4783. ANSWER #1
Cowling and baffles are designed to force air over the cylinder cooling fins. The baffles direct the air close around the cylinders and prevent it from forming hot pools of stagnant air while the main streams rush by unused.
REFERENCE: ITP POWERPLANT, Chapter 7, Pages 27 and 28
             AC65-12A, Pages 314 and 315

4784. ANSWER #3
Cracks at the base of a cooling fin may be acceptable, but it would be necessary to consult the manufacturer's service or overhaul manual to be sure.
REFERENCE: AC65-12A, Page 319

4785. ANSWER #2
After a flight and a few minutes of taxiing, the engine can usually be stopped almost immediately without worrying about excessive temperatures. If the engine is exceptionally hot as indicated by the cylinder-head temperature gage and the oil-temperature gage, it is good practice to allow the engine to idle for a short time before shutdown.
REFERENCE: AIRCRAFT POWERPLANTS, Page 350

4786. ANSWER #3
Cylinder head temperature is usually measured with a thermocouple sensing device. The thermocouple device is usually one of two types; a gasket type which fits under the spark plug, or a bayonet type which fits into the cylinder head.
REFERENCE: ITP POWERPLANT, Chapter 5, Page 30
             AC65-15A, Page 494

4787. ANSWER #1
A double-row radial engine uses a two-throw 180 degree crankshaft to permit the cylinders in each row to be alternately staggered on the common crankcase. That is, the cylinders of the rear row are located directly behind the spaces between the cylinders in the front row. This allows the cylinders in both rows to receive ram air for the necessary cooling.
REFERENCE: AIRCRAFT POWERPLANTS, Page 30

4788. ANSWER #4
With an engine operating near its maximum output, very lean mixtures will cause a loss of power and under certain conditions, serious overheating. When the engine is operated on a lean mixture, the cylinder head temperature gage should be watched closely. An excessively lean mixture can cause extremely high cylinder head temperatures.
REFERENCE: AC65-12A, Page 113

4789. ANSWER #4
When air moves through a turbosupercharger, its temperature is raised because of compression. Because of the problems this heated air could cause, a device called an intercooler is used to cool the air prior to it entering the carburetor. Cool ambient air is the source of cooling.
REFERENCE: ITP POWERPLANT, Chapter 7, Page 13 (Figure 1A-16)
             AC65-12A, Page 81

4790. ANSWER #4
Because of the rich mixture and less than ideal exhaust scavenging present when an engine is idling, prolonged periods of idling will usually result in foreign material building up on the spark plugs.
REFERENCE: AC65-12A, Page 440

4791. ANSWER #2
According to the Aircraft Technical Dictionary, a muff is a shroud placed around a section of the exhaust pipe for the purpose of providing carburetor or cockpit and cabin heat.
REFERENCE: AIRCRAFT TECHNICAL DICTIONARY, Page 156

4792. ANSWER #4
The purpose of a blast tube on an aircraft engine is to cool an engine accessory, such as a generator, or to direct cooling air to a specific point on the engine, like the rear spark plugs.
REFERENCE: AC65-12A, Page 315

4793. ANSWER #1
Of the air moving though a turbine engine, approximately 25% is used to support combustion and the other 75% is used for cooling.
REFERENCE: ITP POWERPLANT, Chapter 2, Pages 44 and 45
             AC65-12A, Pages 50, 51 and 323

4794. ANSWER #3
Part throttle operation of a reciprocating engine reduces volumetric efficiency because it reduces the amount of air the engine is capable of drawing in.
REFERENCE: ITP POWERPLANT, Chapter 1, Page 13
        AC65-12A, Page 38

4795. ANSWER #2
The majority of aircraft engine pistons are machined from aluminum alloy forgings. Grooves are machined in the outside surface to receive the piston rings, and cooling fins are provided on the inside of the piston for greater heat transfer to the engine oil.
REFERENCE: ITP POWERPLANT, Chapter 1, Page 32
        AC65-12A, Page 15

4796. ANSWER #1
Cowl flaps are used to control the operating temperature of recip engines. In flight they are open just enough to maintain the proper operating temperature, because to have them open more than they need to be would increase the drag in flight. During ground operations, the cowl flaps should be opened wide, since drag does not matter.
REFERENCE: AC65-12A, Page 315

4797. ANSWER #3
Increased engine heat will cause the air entering the engine to be heated, and therefore it will be less dense. The less dense the charge entering the engine, the lower the volumetric efficiency.
REFERENCE: ITP POWERPLANT, Chapter 1, Page 13
        AC65-12A, Page 38

4798. ANSWER #3
Inconel steel is used in many exhaust and hot section parts, especially turbine engine hot sections, because of its corrosion resistance and low expansion coefficient. Hot section parts are subjected to numerous impurities present in the exhaust gases and also experience large changes in temperature. The metals used must be able to handle the impurities without corroding and they must be able to handle the heat without expanding too much.
REFERENCE: EA-TEP, Page 44

4799. ANSWER #1
Slip joints are used in recip engine exhaust systems because of the expansion and contraction which takes place in the risers. If the system were built as one solid piece, the differences in expansion and contraction would cause the exhaust system to crack.
REFERENCE: AIRCRAFT POWERPLANTS, Pages 159 and 161 (Figure 7-42)

4800. ANSWER #4
Carburetor air intake heaters normally use the engine exhaust as a source of heat.
REFERENCE: ITP POWERPLANT, Chapter 7, Pages 2 and 3 (Figure 1A-3)

4801. ANSWER #3
One of the most frequent discrepancies that will be detected while inspecting the hot section of a turbine engine is cracking.
REFERENCE: ITP POWERPLANT, Chapter 2, Page 62
        AC65-12A, Page 476

4802. ANSWER #1
Exhaust systems using collector rings make use of telescoping expansion joints (slip joints) to connect the ring to the tailpipe. The slip joints allow the individual units to move independently of each other, and they also make removal, replacement, and alignment of each piece easier.
REFERENCE: AC65-12A, Page 97
        AIRCRAFT POWERPLANTS, Page 160

4803. ANSWER #4
Some intake and exhaust valve stems are hollow and partially filled with metallic sodium. This material is used because it is an excellent heat conductor. The sodium will melt at approximately 208 degrees Fahrenheit, and the reciprocating motion of the valve circulates the liquid sodium and enables it to carry away heat from the valve head to the valve stem, where it is dissipated through the valve guide to the cylinder head.
REFERENCE: ITP POWERPLANT, Chapter 1, Page 30
        AC65-12A, Page 20

4804. ANSWER #1
When repairing or replacing exhaust system components, the proper hardware and clamps should always be used. Steel or low-temperature self-locking nuts should not be substituted for brass or the special high-temperature locknuts used by the manufacturer.
REFERENCE: AC65-12A, Page 101

4805. ANSWER #2
There are two general types of exhaust systems in use on reciprocating aircraft engines: the short stack (open) system and the collector system.
REFERENCE: AC65-12A, Page 96

4806. ANSWER #4
It is generally recommended that exhaust stacks, mufflers, tailpipes, etc., be replaced with new or reconditioned components rather than repaired. Welded repairs to exhaust systems are complicated by the difficulty of accurately identifying the base metal so that the proper repair materials can be selected.
REFERENCE: AC65-12A, Page 101
                      AC43.13-1A, Page 289

4807. ANSWER #1
To satisfy the minimum braking requirements after landing, a thrust reverser should be able to produce in reverse at least 50% of the full forward thrust of which the engine is capable.
REFERENCE: AC65-12A, Page 104

4808. ANSWER #3
An exhaust system that uses a heat exchanger as a source of cabin heat should be inspected by using a hydrostatic test. This involves plugging all openings, pressurizing to 2 psi, and submerging it in water. Any leaks will cause bubbles that can be readily detected.
REFERENCE: AC65-12A, Page 100
                      AC43.13-1A, Page 289

4809. ANSWER #3
Ceramic-coated stacks should be cleaned by degreasing only. They should never be cleaned with sandblast or alkali cleaners.
REFERENCE: AC65-12A, Page 100

4810. ANSWER #2
The exhaust section of the turbojet engine is susceptible to heat cracking. When inspecting the exhaust cone, hotspots may be found. Hotspots are a good indication of a malfunctioning fuel nozzle or combustion chamber.
REFERENCE: AC65-12A, Page 482

4811. ANSWER #4
When welding an exhaust stack, the completed weld should have a smooth seam of uniform thickness and the weld metal should taper smoothly into the base metal.
REFERENCE: AC43.13-1A, Page 30

4812. ANSWER #1
The turbocompound engine consists of a conventional, reciprocating engine in which exhaust-driven turbines are coupled to the engine crankshaft. This system of obtaining additional power is sometimes called a PRT (power recovery turbine) system.
REFERENCE: AC65-12A, Page 89

4813. ANSWER #3
Corrosion-resistant steel parts can be blast-cleaned using sand which has not previously been used on iron or steel. Sand which has previously been used on iron or steel will have particles of those metals trapped in it. These particles, if the sand is used on corrosion-resistant steel, will become imbedded in the parent metal and alter its corrosion resistant qualities. This is the same type of problem which is encountered if galvanized or zinc-plated tools are used, or if the exhaust system is marked on with lead pencil.
REFERENCE: AC65-12A, Page 98

4814. ANSWER #3
The turbocompound engine consists of a conventional, reciprocating engine in which exhaust-driven turbines are coupled to the engine crankshaft. The amount of power the PRT supplies to the engine is determined by the velocity of the exhaust gases that enter it.
REFERENCE: AC65-12A, Page 89

4815. ANSWER #4
Repairs or sloppy weld beads on exhaust components are not acceptable because they can cause local hot-spots and may restrict exhaust gas flow.
REFERENCE: AC65-12A, Page 101

4816. ANSWER #4
Ball joints, on the end of support arms, are sometimes used with large radial engine exhaust systems. The purpose of the ball joint is to let the exhaust move, and the support of course move with it.
REFERENCE: AC65-12A, Pages 97 and 98

4817. ANSWER #3
Galvanized or zinc-plated tools should never be used on the exhaust system, and exhaust system parts should never be marked with a lead pencil. The lead, zinc, or galvanized mark is absorbed by the metal of the exhaust system when heated, creating a distinct change in its molecular structure.
REFERENCE: AC65-12A, Page 98

4818. ANSWER #2
The function of a turbine engine's thrust reverser is to reverse the flow of the exhaust gases. This is done by using either mechanical or aerodynamic blockage reversers.
REFERENCE: ITP POWERPLANT, Chapter 7, Pages 39 and 40
            AC65-12A, Page 103

4819. ANSWER #1
Of the air moving through a turbine engine, approximately 25% is used to support combustion and the other 75% is used for cooling. The air which flows through the combustor to be used for combustion is called primary air. The air which flows around the flame zone and around the combustor for cooling is called secondary air.
REFERENCE: ITP POWERPLANT, Chapter 2, Pages 44 and 45
            AC65-12A, Pages 50, 51 and 323

4820. ANSWER #3
Augmentor tubes are part of the exhaust system. Their function is to create a low pressure zone to help draw cooling air through the engine.
REFERENCE: AC65-12A, Page 101

4821. ANSWER #1
Internal failures in a muffler (baffles, diffusers, etc.) can cause partial or complete engine power loss by restricting the flow of the exhaust gases.
REFERENCE: AC43.13-1A, Pages 285 and 287

4822. ANSWER #1
Engine power loss and excessive backpressure caused by exhaust outlet blockage may be averted by the installation of an exhaust outlet guard. The exhaust outlet guard extends approximately two inches inside the muffler outlet port, thereby preventing any debris from blocking it.
REFERENCE: AC43.13-1A, Page 287

4823. ANSWER #1
Any exhaust system failure should be regarded as a severe hazard. Depending upon the location and type of failure, it can result in carbon monoxide poisoning of the crew and passengers, partial or complete engine power loss, or fire.
REFERENCE: AC43.13-1A, Page 285

4824. ANSWER #4
Prior to any cleaning operation, the exhaust system and surrounding areas should be thoroughly inspected. The cowling and nacelle areas adjacent to the exhaust system should be inspected for telltale signs of exhaust gas soot indicating possible leakage points.
REFERENCE: AC43.13-1A, Page 288

4825. ANSWER #4
The exhaust system of aircraft reciprocating engines often operates at temperatures of 1000 degrees Fahrenheit or more.
REFERENCE: AC43.13-1A, Page 288

4826. ANSWER #4
Approximately one-half of all exhaust system failures are traced to cracks or ruptures in the heat exchanger surfaces used for cabin and carburetor air heat sources. The failures are, for the most part, attributed to thermal and vibration fatigue cracking in areas of stress concentration.
REFERENCE: AC43.13-1A, Page 285

4827. ANSWER #2
A brush block mounted on the engine nose case just behind the propeller contains conductive carbon brushes which transfer electrical power from stationary aircraft components to conductive slip rings which are attached to, and rotate with, the propeller.
REFERENCE: EA-APC, Page 124
            AC65-12A, Page 348

4828. ANSWER #4
The slinger ring is a U-shaped circular channel mounted on the rear of the propeller that incorporates a discharge tube for each propeller blade. Centrifugal force or the slinger ring rotation causes the fluid to flow through the discharge tubes and onto the blades.
REFERENCE: EA-APC, Page 124
            AC65-12A, Page 348

4829. ANSWER #3
Special propeller governers are equipped with magnetic pickups that count propeller revolutions and send a signal to an electronic comparison unit. The comparison unit generates a d.c. pulse and, through electrical actuators, causes a trimming unit to regulate movement of the "slave" governor in order to synchronize it with the "master" propeller governor. On older systems, RPM was sensed by small a.c. generators, but the final adjustment of RPM was accomplished through the propeller governors.
REFERENCE: ITP POWERPLANT, Chapter 8, Page 102
            EA-APC, Page 131
            AC65-12A, Page 347

4830. ANSWER #2
Isopropyl alcohol is generally used because of its availability and low cost. Available at higher cost are phosphate compounds which have the advantage of reduced flammability.
REFERENCE: EA-APC, Page 123
                         AC65-12A, Page 348

4831. ANSWER #2
Differences in speed of two engines under synchronizer control are detected by the propeller governors. One engine and its governor is designated as the "master" and the other as the "slave". When a small (5 to 15) RPM difference in speed is detected, a signal is sent to the "slave" governor and causes it to change the "slave" engine RPM to match that of the "master" engine. These minute RPM changes will go undetected in the cockpit because the tachometers are marked in 100 RPM increments.
REFERENCE: ITP POWERPLANT, Chapter 8, Page 102
                         EA-APC, Page 131
                         EA-AGV, Page 28
                         AC65-12A, Page 347

4832. ANSWER #2
Isopropyl alcohol is the most common fluid used for propeller anti-icing on reciprocating engine aircraft.
REFERENCE: EA-APC, Page 123
                         AC65-12A, Page 348

4833. ANSWER #3
By setting and controlling all propeller governors at exactly the same RPM, excess noise and vibration is eliminated. This is the function of a multi-engine synchronizer system.
REFERENCE: ITP POWERPLANT, Chapter 8, Page 101
                         EA-APC, Page 131
                         EA-AGV, Page 28
                         AC65-12A, Page 347

4834. ANSWER #1
Ice formation destroys the aerodynamic shape of the airfoil, thus reducing thrust. Ice formation also causes an unbalanced condition that induces vibration.
REFERENCE: EA-APC, Page 122
                         AC65-12A, Page 347

4835. ANSWER #4
Ammeters or loadmeters are used to indicate the amount of current drawn by the deicing system.
REFERENCE: EA-APC, Pages 125 and 128
                         AC65-12A, Page 349

4836. ANSWER #2
The anti-icing constant displacement fluid pump is driven by an electric motor. The speed of the motor, and thus the output of the pump, is controlled by a rheostat.
REFERENCE: EA-APC, Page 124
                         AC65-12A, Pages 347 and 348

4837. ANSWER #4
The deicing system components must be checked by means of electrical instruments. On most systems, however, the deicer boot itself is checked by "feel" to determine if it is heating. This method is often used as a preliminary troubleshooting step. Observation of a system meter or loadmeter will not isolate an individual boot as operating or not operating.
REFERENCE: EA-APC, Page 129
                         AC65-12A, Page 349

4838. ANSWER #4
Propeller manufacturers publish the proper lubrication procedures, complete with oil and grease specifications.
REFERENCE: ITP POWERPLANT, Chapter 8, Page 76
                         EA-APC, Page 52
                         AC65-12A, Page 354

4839. ANSWER #3
See the explanation for #4838.
REFERENCE: ITP POWERPLANT, Chapter 8, Page 76
                         EA-APC, Page 52
                         AC65-12A, Page 354

4840. ANSWER #1
Plasticity is a state or condition of being easily subjected to molding into any form under pressure. It means that the grease should not become gummy or hard at low temperatures but should flow easily.
REFERENCE: AIRCRAFT POWERPLANTS, Page 174

4841. ANSWER #1
When a propeller is placed in a vertical position on a balancing stand and it persists in moving to a horizontal position, it is said to be out of static vertical balance. A small plate is sometimes attached to the light side of the propeller boss to correct this problem.
REFERENCE: EA-APC, Page 20
              AC65-12A, Pages 352 and 353 (Figure 7-29)

4842. ANSWER #1
The arbor is inserted into the propeller bore. It supports and allows the free rotation of the propeller between the hardened steel knife edges of the balancing stand.
REFERENCE: AC65-12A, Page 353

4843. ANSWER #4
All blades of a propeller must be of equal dimension for balance purposes. Vibration and subsequent blade failure can occur if a propeller is not in static, dynamic and aerodynamic balance. If the shape or length of one blade is changed, all the blades must be altered to the same dimensions.
REFERENCE: AC43.13-1A, Pages 223 and 229

4844. ANSWER #1
When a separate metal hub is used, it becomes a component part of the rotating propeller assembly. The metal hub must be included in the final balancing process.
REFERENCE: EA-APC, Page 29
              AC65-12A, Page 353

4845. ANSWER #4
Vibration resulting from a propeller unbalanced condition increases with the speed of rotation. Increased centrifugal force results in an increased vibration.
REFERENCE: EA-APC, Pages 14 and 15
              AC65-12A, Page 350

4846. ANSWER #4
An amount of solder necessary to correct the unbalanced condition is melted and smoothed onto the face side of the metal tipping cap of the light blade.
REFERENCE: EA-APC, Page 20
              AC43.13-1A, Page 222

4847. ANSWER #1
With speeder spring tension reduced, the governor flyweights will tilt outward into an overspeed position. Sensing the overspeed, the governor will port oil in the direction necessary to increase propeller blade angle. The blade angle increasing will cause a reduction of RPM, which will cause a manifold pressure increase if the throttle position remains the same. In this situation, the propeller control has been moved toward the decrease RPM position.
REFERENCE: ITP POWERPLANT, Chapter 8, Page 65
              EA-APC, Page 54
              AC65-12A, Page 335

4848. ANSWER #1
The pulley stop screw actually limits the amount of tension that can be put on the governor speeder spring and thereby limits engine RPM by propeller blade angle control. Final adjustment to maximum RPM for takeoff is made with this stop screw.
REFERENCE: ITP POWERPLANT, Chapter 8, Page 64
              EA-APC, Page 63
              AC65-12A, Pages 346 and 347 (Figure 7-24)

4849. ANSWER #4
In this circumstance the governor senses an underspeed condition. Oil is ported to position the blade angle in the full low pitch position in an effort to reduce engine load and increase RPM. The propeller now functions as a fixed-pitch propeller at the low blade angle, high RPM setting.
REFERENCE: ITP POWERPLANT, Chapter 8, Page 65
              EA-APC, Page 54
              AC65-12A, Page 340

4850. ANSWER #3
With a power increase, the governor senses an overspeed condition. Oil is ported to increase the blade angle so that the RPM can be maintained at the higher power setting and the efficient low angle of attack can be maintained. Research in propeller aerodynamics has proven that a low (2 to 4 degree) angle of attack is most efficient in all forward airspeed and RPM situations.
REFERENCE: ITP POWERPLANT, Chapter 8, Page 67
              EA-APC, Page 54
              AC65-12A, Page 335

4851. ANSWER #1
The propeller governor controls the oil into and out of the pitch change mechanism. In this manner the blade angle (the load on the engine) and thus the RPM of the engine is controlled.
REFERENCE: ITP POWERPLANT, Chapter 8, Page 64
                        EA-APC, Page 53
                        AC65-12A, Page 335

4852. ANSWER #4
When governor flyweight and speeder spring forces are in equilibrium, no oil flows to or from the pitch change mechanism. There is no blade angle or RPM change when this is happening, so it is called an "on-speed" condition.
REFERENCE: ITP POWERPLANT, Chapter 8, Page 65
                        EA-APC, Page 54
                        AC65-12A, Page 355 (Figure 7-13)

4853. ANSWER #2
When the governor flyweights tilt outward, the pilot valve is raised to an "overspeed" position. When the flyweights tilt inward, the pilot valve is lowered to an "underspeed" position. Flyweight movement is opposed by tension on the speeder spring.
REFERENCE: ITP POWERPLANT, Chapter 8, Page 65
                        EA-APC, Page 54
                        AC65-12A, Page 335

4854. ANSWER #4
See the explanation to question #4853.
REFERENCE: ITP POWERPLANT, Chapter 8, Page 65
                        EA-APC, Page 54
                        AC65-12A, Page 335

4855. ANSWER #1
When the propeller control in the cockpit is actuated, the compression on the speeder spring is changed. More compression on the speeder spring results in higher engine RPM. It should be noted that the word compression is a better descriptor than the word tension.
REFERENCE: ITP POWERPLANT, Chapter 8, Page 65
                        EA-APC, Page 54
                        AC65-12A, Page 335

4856. ANSWER #4
When the propeller control in the cockpit is moved forward, governor speeder spring compression is increased. A governor "under-speed" condition is created. The pilot valve is positioned to port oil in the direction necessary to decrease propeller blade angle, thereby increasing RPM.
REFERENCE: ITP POWERPLANT, Chapter 8, Page 65
                        EA-APC, Page 54
                        AC65-12A, Page 335

4857. ANSWER #4
The speeder spring provides the compression load against the flyweights in the governor. Speeder spring force is controlled by the cockpit propeller lever.
REFERENCE: ITP POWERPLANT, Chapter 8, Page 65
                        EA-APC, Page 54
                        AC65-12A, Page 340

4858. ANSWER #4
See the explanation to question #4847.
REFERENCE: ITP POWERPLANT, Chapter 8, Page 65
                        EA-APC, Page 54
                        AC65-12A, Page 335

4859. ANSWER #2
The greatest stress on a propeller is the centrifugal force that is created by the propeller's rotation. Centrifugal force may be greater than 25 tons, depending on the blade weight and RPM.
REFERENCE: ITP POWERPLANT, Chapter 8, Page 42
                        EA-APC, Page 13
                        AC65-12A, Page 327

4860. ANSWER #2
Aerodynamic twisting force tends to twist a propeller blade to a higher angle by aerodynamic action. This results, in part, from the fact that the center of lift (also called center of pressure) is toward the leading edge of the blade.
REFERENCE: ITP POWERPLANT, Chapter 8, Page 42
                        EA-APC, Page 14
                        AC65-12A, Pages 327 and 328

4861. ANSWER #4
In the ground operational range on a turbo-propeller installation, the propeller blade angle is not controlled by the propeller governor, but rather by the power lever position.
REFERENCE: ITP POWERPLANT, Chapter 8, Page 91
                    EA-APC, Page 99
                    AC65-12A, Page 354

4862. ANSWER #2
See the explanation to question #4860.
REFERENCE: ITP POWERPLANT, Chapter 8, Page 42
                    EA-APC, Page 14
                    AC65-12A, Pages 327 and 328

4863. ANSWER #1
High RPM position is the low pitch position. When a propeller moves to reverse pitch, the blades rotate below the low blade angle and directly into a negative angle of about −15 degrees. On a hydromatic propeller, the low pitch stop is retracted in order for the blades to move to the reverse pitch setting.
REFERENCE: ITP POWERPLANT, Chapter 8, Page 35
                    EA-APC, Page 96
                    AC65-12A, Page 331

4864. ANSWER #3
If a propeller has been subjected to salt water, it should be flushed with fresh water until all traces of salt have been removed. This should be accomplished as soon as possible after the salt water has splashed on the propeller. After flushing, all parts should be dried thoroughly, and metal parts should be coated with clean engine oil or a suitable equivalent.
REFERENCE: AC65-12A, Page 353

4865. ANSWER #2
Magnafluxing is an accepted method for non-destructive testing of steel propellers.
REFERENCE: ITP POWERPLANT, Chapter 8, Page 51
                    EA-APC, Page 28
                    AC65-12A, Page 350

4866. ANSWER #1
All forces acting on the propeller are concentrated on the shank, and any damage in this area is critical.
REFERENCE: ITP POWERPLANT, Chapter 8, Page 47
                    EA-APC, Page 47
                    AC43.13-1A, Page 229

4867. ANSWER #1
Pitch distribution from hub to tip requires that a specific blade station must be known in order to measure blade angle. A reference blade angle measuring station is always specified by the propeller manufacturer.
REFERENCE: ITP POWERPLANT, Chapter 8, Page 57
                    EA-APC, Page 3
                    AC65-12A, Pages 351 and 352

4868. ANSWER #1
The propeller blade angle is the acute angle between the air-foil section chord line (at the reference station) and the propeller rotational plane. The reference station is often at a point that is 75% of the distance from the hub to the tip.
REFERENCE: ITP POWERPLANT, Chapter 8, Page 34
                    EA-APC, Page 2
                    AC65-12A, Page 326

4869. ANSWER #3
High-speed, high-altitude cruising flight requires a high blade angle which is also a high pitch. This means a greater distance of forward movement with each revolution of the propeller. The other flight conditions in this question would require a low pitch setting.
REFERENCE: ITP POWERPLANT, Chapter 8, Page 35
                    EA-APC, Page 46
                    AC65-12A, Page 329

4870. ANSWER #4
The situation in this question indicates that the pressure cutout switch is stuck in the closed position and is not performing its major function, that of opening the circuit to the feather pump motor to shut it off. The pump continues to cause an oil pressure rise, which causes the distributor valve to shift and unfeather the propeller.
REFERENCE: ITP POWERPLANT, Chapter 8, Page 87
                    EA-APC, Page 89
                    AC65-12A, Page 343

4871. ANSWER #1
Effective pitch is the actual distance that a propeller moves through the air in one revolution. Geometric pitch is a theoretical distance of movement in one revolution. The difference between the two types of pitch is known as slip and indicates the propeller efficiency.
REFERENCE: ITP POWERPLANT, Chapter 8, Page 43
                        EA-APC, Page15
                        AC65-12A, Page 325

4872. ANSWER #1
The pitch-changing oil lubricates the hydromatic propeller. No other lubrication is required for the pitch-change mechanism.
REFERENCE: ITP POWERPLANT, Chapter 8, Page 89
                        EA-APC, Page 94
                        AC65-12A, Page 354

4873. ANSWER #1
When the throttle is opened on a recip engine which has a constant-speed propeller, operating in the constant-speed range, the propeller control system will cause the blade angle to increase in order to absorb the additional power and maintain the same RPM.
REFERENCE: ITP POWERPLANT, Chapter 8, Page 67
                        EA-APC, Page 54
                        AC65-12A, Page 330

4874. ANSWER #2
Propeller blade stations are measured from the hub centerline to the tip of each blade. Each blade has its own "one inch" stations starting from station "0" at the hub centerline.
REFERENCE: ITP POWERPLANT, Chapter 8, Page 40
                        EA-APC, Page 12
                        AC65-12A, Pages 325 and 326

4875. ANSWER #3
A propeller is a rotating airfoil and creates thrust the same way an airplane's wing creates lift. A low pressure area is formed on the front of the blade, and the propeller moves forward into this decreased pressure area.
REFERENCE: ITP POWERPLANT, Chapter 8, Page 40
                        EA-APC, Page 12
                        AC65-12A, Page 327

4876. ANSWER #1
Stopping a Hamilton Standard counterweight propeller in full high pitch covers the piston surface with the cylinder, which acts to prevent corrosion and dirt accumulation on the piston. Additional reasons for this action are that it prevents congealing of the oil in the cylinder when operating in cold climates. It also prevents oil from going to the propeller cylinder rather than to the engine bearings when the engine is first started. Propellers without this design feature should be stopped in low pitch to reduce engine load for starting.
REFERENCE: ITP POWERPLANT, Chapter 8, Page 60
                        EA-APC, Page 48

4877. ANSWER #1
The low pitch stop on a constant-speed propeller is set so that the engine can develop rated power at sea level with the RPM specified by the manufacturer. A low pitch stop improperly set can result in engine overspeed or failure to attain rated power.
REFERENCE: ITP POWERPLANT, Chapter 8, Page 36
                        EA-APC, Page 8

4878. ANSWER #3
Propeller angle-of-attack is the acute angle between the blade chord and the relative wind. The angle of attack on a fixed pitch propeller can change with different flight conditions since relative wind is a result of the combined velocities of RPM (rotational velocity) and air speed (forward velocity).
REFERENCE: ITP POWERPLANT, Chapter 8, Pages 40 and 41
                        EA-APC, Page 13
                        AC65-12A, Page 327

4879. ANSWER #2
Centrifugal twisting moment tries to force the blades toward a low blade angle. It is a greater force than aerodynamic twisting moment, which opposes it. See questions #4860 and #4862.
REFERENCE: ITP POWERPLANT, Chapter 8, Page 42
                        EA-APC, Page 14
                        AC65-12A, Page 327

4880. ANSWER #1
The curved or cambered side of a propeller blade is called the blade back. The flat side of the blade is called the face or thrust side.
REFERENCE: ITP POWERPLANT, Chapter 8, Page 34
                       EA-APC, Page 2
                       AC65-12A, Page 326

4881. ANSWER #3
The low RPM position of the propeller is the high pitch position. The blade moves from high pitch directly to the feather position.
REFERENCE: ITP POWERPLANT, Chapter 8, Page 76
                       EA-APC, Page 74
                       AC65-12A, Page 335

4882. ANSWER #4
The feather button holding coil holds a feather relay closed that applies power to the propeller feather motor, which drives the feather pump.
REFERENCE: ITP POWERPLANT, Chapter 8, Page 87
                       EA-APC, Pages 88 and 89
                       AC65-12A, Page 343

4883. ANSWER #2
Metal tipping is applied to the leading edge of most wooden propeller blades to prevent damage from small stones and debris which might strike the prop during ground operations. This tipping is attached to the blade with counter sunk screws in the thick blade sections and with copper rivets in the thin sections.
REFERENCE: ITP POWERPLANT, Chapter 8, Page 44
                       EA-APC, Page 18
                       AC65-12A, Page 332

4884. ANSWER #3
Blade angle is the acute angle formed by a line perpendicular to the crankshaft centerline, which is called the plane of rotation, and the chord of the blade at a specified reference station.
REFERENCE: ITP POWERPLANT, Chapter 8, Page 34
                       EA-APC, Page 2
                       AC65-12A, Page 326

4885. ANSWER #1
Propeller blade stations increase from the hub center line, which would be station "0", to the tip of the blade.
REFERENCE: ITP POWERPLANT, Chapter 8, Page 40
                       EA-APC, Page 12
                       AC65-12A, Pages 325 and 326

4886. ANSWER #2
Aerodynamic (twisting) forces on a propeller blade tend to increase its pitch.
REFERENCE: ITP POWERPLANT, Chapter 8, Page 42
                       EA-APC, Page 14
                       AC65-12A, Page 327

4887. ANSWER #4
The pitch change gear preload (lash) on Hamilton Standard hydromatic propellers is adjusted by shim thickness between the dome barrel shelf and the stationary cam base. Improper adjustment can result in erratic or jerky operation of the propeller.
REFERENCE: ITP POWERPLANT, Chapter 8, Page 89
                       EA-APC, Page 94

4888. ANSWER #3
Centrifugal force acting on the counterweights of nearly all counterweight-type propellers operate to move the blades toward a high pitch position.
REFERENCE: ITP POWERPLANT, Chapter 8, Page 60
                       EA-APC, Page 61
                       AC65-12A, Page 333

4889. ANSWER #1
The hydromatic propeller distributor valve shifts only for the unfeather operation. During all constant-speed and feather operations, the distributor valve keeps engine oil pressure and governor oil pressure separated. Its function is to direct these different levels of oil pressure to opposite sides of the propeller pistion.
REFERENCE: ITP POWERPLANT, Chapter 8, Page 88
                EA-APC, Pages 89 and 90
                AC65-12A, Page 339

4890. ANSWER #4
The high RPM position is the low pitch position. Therefore, the blade must move from low pitch through high pitch to get to the feather position.
REFERENCE: ITP POWERPLANT, Chapter 8, Page 76
                EA-APC, Page 74
                AC65-12A, Page 335

4891. ANSWER #3
Solid aluminum propeller blades should be routinely washed with soap and water, rinsed with fresh water, then dried and lightly oiled. This is a good corrosion preventive measure and also gives the opportunity for a close inspection for nicks, dents, or other damage that could exist under a coating of dust or dirt.
REFERENCE: ITP POWERPLANT, Chapter 8, Page 47
                EA-APC, Page 21
                AC65-12A, Page 353

4892. ANSWER #3
A propellers's blade angle decreases along its length from hub to tip. Also, a propeller blade airfoil shape changes from a low speed airfoil near the hub to a high speed shape at the tip. This design feature is necessary because the tip of a propeller travels considerably faster than its hub. This twist of the blade is termed pitch distribution.
REFERENCE: ITP POWERPLANT, Chapter 8, Page 40
                EA-APC, Page 12
                AC65-12A, Page 326

4893. ANSWER #1
The thinner the blade section, the greater the amount the blade can be cold straightened. The straightening of a bent blade is classified as a major repair and must be accomplished by a certified propeller repair station or by the propeller manufacturer.
REFERENCE: ITP POWERPLANT, Chapter 8, Page 48 (Figure 3B-6)
                EA-APC, Pages 22 and 23 (Figure 39)
                AC43.13-1A, Page 229

4894. ANSWER #1
This question describes a normally functioning Hamilton Standard hydromatic propeller. When the feather button is pushed in, a holding coil, which gets an electrical ground from the pressure cutout switch on the propeller governor, keeps the button depressed. When the feather cycle is complete, oil pressure increases until the switch on the governor is opened, causing the holding coil to lose its ground contact and open. For unfeathering the same circuitry is used. However, the button is pushed and held in to prevent the button from popping back out when the pressure cutout switch opens. The feathering pump builds pressure higher than the cutout switch setting and causes the distributor valve to shift, which redirects the oil flow to the opposite side of the propeller piston. Piston movement in the opposite direction accomplishes the unfeathering operation.
REFERENCE: ITP POWERPLANT, Chapter 8, Pages 87 and 88
                EA-APC, Page 89
                AC65-12A, Page 343

4895. ANSWER #1
The etching process, utilizing a caustic soda solution and a nitric acid neutralizer, is used in much the same manner as a dye penetrant inspection. The AC43.13-1A gives a detailed description of the etching process.
REFERENCE: AC43.13-1A, Pages 232 and 233

4896. ANSWER #4
The limits of a constant-speed propeller range are controlled by the mechanical stops of the propeller. Within this range, RPM is controlled by the propeller governor setting. If the governor should fail, the low blade angle stop setting must allow a static RPM of no more than 103% of rated RPM (FAR 23.33 and FAR 25.33).
REFERENCE: ITP POWERPLANT, Chapter 8, Page 63
                       EA-APC, Page 53
                       AC65-12A, Page 330

4897. ANSWER #4
To allow the engine to develop its rated takeoff power, the constant-speed propeller is normally set in the low pitch, high RPM position.
REFERENCE: ITP POWERPLANT, Chapter 8, Page 59
                       EA-APC, Page 46
                       AC65-12A, Page 328

4898. ANSWER #3
Hamilton Standard counterweight propellers have brackets attached to the blade butts. The travel of these brackets, and thus the travel of the blades, is restricted by high and low pitch stops located in the counterweights.
REFERENCE: ITP POWERPLANT, Chapter 8, Page 60
                       EA-APC, Pages 48 and 59

4899. ANSWER #1
The two-position and the constant-speed counterweight propellers both use hydraulic force to decrease blade angle and centrifugal force acting on counterweights to increase blade angle. The major difference is that the constant-speed propeller utilizes a governor to boost the oil pressure to a higher level and automatically controls the oil flow to and from the propeller. The two-position propeller operates at engine lubrication system pressure with the oil flow controlled with a manual selector valve.
REFERENCE: ITP POWERPLANT, Chapter 8, Page 59
                       EA-APC, Page 47

4900. ANSWER #1
Type Certificate Data Sheets for engine-propeller combinations identify any critical RPM ranges that are to be avoided to prevent severe vibration. Regulations require that these ranges be marked on the tachometer with a red arc.
REFERENCE: ITP POWERPLANT, Chapter 8, Page 36
                       EA-APC, Page 8

4901. ANSWER #2
Oversize or elongated bolt holes on a wooden propeller can be cause for rejection. However, some oversize or worn bolt holes may be repaired at a repair station by the use of metal inserts to restore the original diameter. The other defects described in this question are considered minor and can be repaired by a certified powerplant mechanic.
REFERENCE: ITP POWERPLANT, Chapter 8, Page 46
                       EA-APC, Page 20
                       AC65-12A, Page 220

4902. ANSWER #3
Thrust washers become a permanent part of a Hamilton Standard blade assembly because they are installed on the blade shank before the blade butt flange is forged. A special thrust type roller bearing is installed between the thrust washer and an interior bearing surface of the hub. During operation, the thrust washer bears the tremendous tension load caused by centrifugal force. The bearing allows the blade to turn under the load.
REFERENCE: EA-APC, Pages 84 and 85 (Figures 149 and 150)

4903. ANSWER #4
Bending and torsion stresses are unavoidable during propeller operation. These stresses, combined with nicks on an aluminum alloy blade, can lead to fatigue cracks and eventual blade separation.
REFERENCE: ITP POWERPLANT, Chapter 8, Page 47
                       EA-APC, Page 21

4904. ANSWER #3
Major repairs to all propellers and blades must be done by an appropriately certificated propeller repair station or the propeller manufacturer.
REFERENCE: ITP POWERPLANT, Chapter 8, Page 39
                       EA-APC, Page 11

4905. ANSWER #4
A blade cuff is a metal, wood or plastic structure designed for attachment to the shank of the propeller blade. The cuff surface transforms the round shank into an airfoil section. The cuff is designed primarily to increase the flow of cooling air to the engine nacelle.
REFERENCE: AC65-12A, Page 358

4906. ANSWER #1
A three-way propeller valve is a selector valve used in a two-position propeller control system. It is used to direct oil at engine lubrication system pressure to the propeller or drain the oil from the propeller and return it to the engine sump.
REFERENCE: ITP POWERPLANT, Chapter 8, Page 59
             EA-APC, Page 47 (Figure 83)

4907. ANSWER #3
The purpose of a propeller is to convert engine horsepower to useful thrust. Modern propellers may convert as much as 85% of the engine brake horsepower to thrust horsepower. This is known as propulsive efficiency.
REFERENCE: ITP POWERPLANT, Chapter 8, Page 33
             EA-APC, Page 2
             AC65-12A, Page 325

4908. ANSWER #2
The constant-speed propeller provides maximum efficiency by adjusting the blade angle in flight to maintain a constant RPM. This action enables the blade chord to move with the relative wind, thus maintaining a constant low and efficient angle of attack.
REFERENCE: ITP POWERPLANT, Chapter 8, Page 64
             EA-APC, Page 53
             AC65-12A, Page 327

4909. ANSWER #4
The centrifugal twisting force, sometimes called centrifugal twisting moment or CTM, is greater than the aerodynamic twisting force and tries to force the blades to a low blade angle.
REFERENCE: ITP POWERPLANT, Chapter 8, Page 43
             EA-APC, Page 14
             AC65-12A, Page 327

4910. ANSWER #2
Geometric pitch of a propeller is the theoretical distance that the propeller will move forward in one revolution. Effective pitch is the actual distance that the propeller does move forward in one revolution. The difference between geometric pitch and effective pitch is called slippage. Therefore, effective pitch plus slippage is equal to geometric pitch.
REFERENCE: ITP POWERPLANT, Chapter 8, Page 44 (Figure 2B-11)
             EA-APC, Page 15 (Figure 27)
             AC65-12A, Page 326 (Figure 7-1)

4911. ANSWER #3
Propeller blade angle is the acute angle formed by the chord line of the blade and the rotational plane of the propeller. The rotational plane is always perpendicular to the centerline of the engine crankshaft. The airframe longitudinal centerline has no effect on the measurement of propeller blade angle.
REFERENCE: ITP POWERPLANT, Chapter 8, Page 34
             EA-APC, Page 2
             AC65-12A, Page 326

4912. ANSWER #4
Torque bending force, in the form of air resistance, tends to bend the propeller blades opposite to the direction of rotation.
REFERENCE: ITP POWERPLANT, Chapter 8, Page 42
             EA-APC, Page 14
             AC65-12A, Page 327

4913. ANSWER #4
Thrust bending force is the force that tends to bend the propeller tips forward. This force is comparable to the coning action of a helicopter rotor blade, except that the thrust bending force acts forward instead of upward.
REFERENCE: ITP POWERPLANT, Chapter 8, Page 42
             EA-APC, Page 13
             AC65-12A, Page 327

4914. ANSWER #3
During takeoff, a constant-speed propeller requires high-speed and low-pitch angle in order to allow the engine to develop maximum rated power.
REFERENCE: ITP POWERPLANT, Chapter 8, Page 59
             EA-APC, Page 46
             AC65-12A, Page 328

4915. ANSWER #4
Moisture condenses between the metal tipping and the wood. The tipping is provided with small holes near the blade tip to allow this moisture to be thrown out by centrifugal force. It is important that these drain holes be kept open at all times. More moisture in one blade than another can cause enough unbalance to cause propeller vibration.
REFERENCE: ITP POWERPLANT, Chapter 8, Page 44
EA-APC, Page 18
AC65-12A, Page 332

4916. ANSWER #3
During the first run of a newly installed hydromatic propeller it is necessary to replace the air trapped in the dome with pitch-change oil. This is accomplished by several full travel movements of the propeller piston, which forces the air back to the engine sump and case breather. Little if any air will be trapped in the dome if it is filled with oil before the dome seal nut is installed.
REFERENCE: AIRCRAFT POWERPLANTS, Page 393

4917. ANSWER #4
During installation of a spline shaft propeller, front cone bottoming occurs if the apex of the front cone is allowed to contact the ends of the shaft splines. In this case, neither front nor rear cone can be tightened into the cone seats of the propeller hub and the hub will be loose on the shaft.
REFERENCE: ITP POWERPLANT, Chapter 8, Page 53
EA-APC, Pages 31 and 32

4918. ANSWER #4
A damaged piston-to-dome seal in a hydromatic propeller will result in sluggish operation of the pitch change mechanism. The effect is the same as with any hydraulic actuator that has a damaged piston seal.
REFERENCE: ITP POWERPLANT, Chapter 8, Page 89
EA-APC, Page 94

4919. ANSWER #2
The blades of a hydromatic propeller should be in the full-feathered position when the dome assembly is installed. Since the pitch-change mechanism, with the feather and low blade angle stops, is contained in the dome assembly, both the dome and the blades must be in the full-feather position to ensure their relative positions when the gears are meshed.
REFERENCE: ITP POWERPLANT, Chapter 8, Page 88
EA-APC, Page 92

4920. ANSWER #4
If front cone bottoming is indicated when installing a spline shaft propeller, a spacer should be installed behind the rear cone to move the entire hub assembly forward over the ends of the shaft splines. When this is accomplished, the apex of the front cone can no longer contact the shaft splines and tightening can be accomplished.
REFERENCE: ITP POWERPLANT, Chapter 8, Page 53
EA-APC, Page 32

4921. ANSWER #2
When the full feathered position has been reached on a hydromatic propeller, the oil pressure delivery is normally stopped automatically by the opening of an oil pressure operated, electric cutout switch. This stops the electrically driven feather pump.
REFERENCE: ITP POWERPLANT, Chapter 8, Page 87
EA-APC, Page 89
AC65-12A, Page 343

4922. ANSWER #3
On a hydromatic propeller, incorrect preload on the beveled rotating cam gear in the dome and the gear segments on the blade butts can cause excessive binding or backlash of the gear teeth.
REFERENCE: ITP POWERPLANT, Chapter 8, Page 89
EA-APC, Page 94

4923. ANSWER #1
The purpose of the cones for a spline shaft propeller installation is to support and align the hub on the shaft. This is similar to the action of tapered bearings which position and support a wheel on an axle.
REFERENCE: ITP POWERPLANT, Chapter 8, Page 52
EA-APC, Page 30
AC65-12A, Page 332

4924. ANSWER #4
AC43.13-1A specifies that when a fixed pitch wooden propeller has been installed, the bolts should be checked for tightness after the first flight, after the first 25 hours of flying, and at least every 50 flying hours thereafter. The concern here is the moisture content of the wood fibers. The fibers can shrink from engine heat, causing the bolts to become loose. The fibers can also swell from humidity, which will cause the bolts to be too tight.
REFERENCE: AC43.13-1A, Page 213

4925. ANSWER #3
A loose retaining nut on a spline shaft propeller installation can cause galling and wear to both front and rear cones and the cone seat on the hub. Propeller vibration could be an indication of a loose retaining nut.
REFERENCE: ITP POWERPLANT, Chapter 8, Pages 52 to 56
EA-APC, Pages 30 to 32

4926. ANSWER #1
Hydraulically operated constant-speed propellers should be placed in the high RPM, low pitch position for all ignition and magneto checking. This causes the propeller to operate as a fixed-pitch propeller and provides an RPM standard for determining the operating condition of the engine.
REFERENCE: AC65-12A, Page 437

4927. ANSWER #3
On a hydromatic propeller, the spider shaft oil seal is utilized to prevent pitch change oil from leaking between the spider and the propeller shaft and out around the rear cone.
REFERENCE: EA-APC, Page 85 (Figure 150)

4928. ANSWER #2
Bearing blue color transfer, sometimes called prussian blue, is used to determine the amount of surface contact between a tapered propeller shaft and the propeller hub. At least 70% surface contact is required.
REFERENCE: ITP POWERPLANT, Chapter 8, Page 51
EA-APC, Page 28

4929. ANSWER #4
Propeller blade tracking is the process of determining the blade tip positions relative to each other. A propeller out-of-track condition may indicate a bent propeller shaft or a blade that is bent.
REFERENCE: ITP POWERPLANT, Chapter 8, Pages 53 and 54
EA-APC, Page 32
AC65-12A, Page 350

4930. ANSWER #3
The three small holes (No. 60 drill) in the metal tipping of a wooden propeller serve to ventilate and release moisture formed by condensation between the tipping and the wooden blade.
REFERENCE: ITP POWERPLANT, Chapter 8, Page 44
EA-APC, Page 18
AC65-12A, Page 332

4931. ANSWER #4
Correction of an out-of-track condition on a fixed pitch wooden propeller may be made by inserting paper or brass shims between the inner flange of the separate metal hub and the propeller boss. On flange type shaft propeller installations, the shim is placed between the propeller boss and the propeller shaft flange.
REFERENCE: ITP POWERPLANT, Chapter 8, Page 54
EA-APC, Page 33

4932. ANSWER #3
To manually feather a hydromechanical propeller, such as is used on a light twin engine aircraft, the propeller control lever is pulled into the feather position. This action ports the governor oil pressure from the cylinder of the propeller and allows the force of springs and counterweights or compressed air to drive the propeller to the feather blade angle.
REFERENCE: ITP POWERPLANT, Chapter 8, Page 79
EA-APC, Page 77
AC65-12A, Page 255

4933. ANSWER #3
Hydraulically operated constant-speed propellers should be placed in the high RPM, low pitch position for all ignition and magneto checking. This causes the propeller to operate as a fixed-pitch propeller and provides an RPM standard for determining the operating condition of the engine.
REFERENCE: AC65-12A, Page 437

4934. ANSWER #3
On a hydromatic, full feathering propeller, the electric driven propeller feathering pump is shut off at termination of the feather operation by the opening of an oil pressure operated, electric cut off switch.
REFERENCE: ITP POWERPLANT, Chapter 8, Page 87
EA-APC, Page 89
AC65-12A, Page 343